D0926930

THE KILL CHAIN

THE KILL CHAIN

DEFENDING AMERICA IN THE FUTURE OF HIGH-TECH WARFARE

CHRISTIAN BROSE

NEW YORK

Hachette Books
Hachette Book Group
1290 Avenue of the Americas
New York, NY 10104
HachetteBooks.com
Twitter.com/HachetteBooks
Instagram.com/HachetteBooks

First Edition: April 2020

Hachette Books is a division of Hachette Book Group, Inc.
The Hachette Books name and logo are trademarks of Hachette Book Group, Inc.

The publisher is not responsible for websites (or their content) that are not owned by the publisher.

The Hachette Speakers Bureau provides a wide range of authors for speaking events. To find out more, go to www.hachettespeakersbureau.com or call (866) 376-6591.

Library of Congress Cataloging-in-Publication Data
Names: Brose, Christian, author.
Title: The kill chain: defending America in the future of high-tech warfare / Christian Brose.
Other titles: Defending America in the future of high-tech warfare
Identifiers: LCCN 2019056319 | ISBN 9780316533539 (hardcover) | ISBN 9780316533362 (ebook)
Subjects: LCSH: United States—Military policy—21st century. | Military art and science—Technological innovations. | United States—Strategic aspects. | China—Strategic aspects. | Access denial (Military science) | Weapons systems—United States. | Russia (Federation)—Strategic aspects. | United States—Defenses. | Strategy.
Classification: LCC UA23 .B7838 2020 | DDC 355/.033573—dc23
LC record available at https://lccn.loc.gov/2019056319

Printed in the United States of America

LSC-C

10 9 8 7 6 5 4 3 2 1

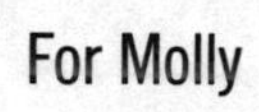

For Molly

CONTENTS

INTRODUCTION

PLAYING A LOSING GAME

One of the last conversations I ever had with John McCain in person was in the winter of 2017, shortly before he left Washington for the last time. We talked about how the United States could lose a war with China—not in the distant future, but now.

For most of the prior decade, I had been McCain's principal advisor on national security and military issues. During the last four years of his life, when he was chairman of the Senate Armed Services Committee, I was his staff director. That meant I led a team of defense policy experts who supported McCain and his colleagues in authorizing and overseeing the entire US defense program—every policy and activity of the Department of Defense, every weapon it developed and bought, every dollar of the roughly $700 billion that it spent each year. McCain and I had access to the Pentagon's most highly classified secrets and programs, and we regularly met with our nation's top defense officials and highest-ranking military officers.

That is what we had just finished doing on that winter day in 2017. McCain had directed me to set up a briefing for all one hundred US senators about the problem that had haunted us and motivated our work together for the past several years: the accelerating erosion of the US military's technological advantage over other great powers, primarily China, which was rapidly building up arsenals of advanced weapons with the explicit purpose of being able to fight and win a war against the United States. McCain wanted his fellow senators to know that America was falling behind and at risk of losing a race that most of them did not even know was being run.

For years, McCain and I had been pleading with Pentagon leaders to be clearer and more forthcoming with Congress and the American people about how bad things really were. They did not want to encourage our competitors by sounding defeatist, which was an apt concern. But it was a concern we had to overcome because it is impossible to solve a problem that no one knows exists. As it stood, the Chinese Communist Party knew far more about the US military and its vulnerabilities than the American people and their elected representatives did.

That year, things seemed like they were starting to change. The chairman of the Joint Chiefs of Staff, General Joseph Dunford, had testified to McCain's committee in June. "In just a few years," he said, "if we do not change our trajectory, we will lose our qualitative and quantitative competitive advantage."[1] In other words, the US military would no longer be the best.

A few months later, the RAND Corporation, a renowned nonpartisan research institute whose military analysis McCain and I consumed regularly, concluded in a major report that "U.S. forces could, under plausible assumptions, lose the next war they are called upon to fight."[2]

That assessment was echoed by a bipartisan commission of military experts that McCain had established through legislation that year to provide an independent examination of US defense strategy. They rendered their judgment to Congress shortly after McCain's death in 2018. "America's military superiority...has eroded to a dangerous degree," they wrote. "The U.S. military could suffer unacceptably high casualties and loss of major capital assets in its next conflict. It might struggle to win, or perhaps lose, a war against China or Russia."[3]

McCain wanted the briefing that day to be a wake-up call to his colleagues—to provide many of the details behind these startling public pronouncements and to build greater support for the new technologies, ideas, reforms, and resources that McCain and I had been trying for years to champion. All ninety-nine of McCain's Senate colleagues were invited. About a dozen showed up.

For those senators who were there, it was a depressing dose of reality. The person who provided the briefing that day was a former Pentagon official in the Obama administration named David Ochmanek. A year later, he spoke publicly about the many war games—what are essentially simulations of future wars—that he has conducted for the Department of Defense upon leaving government. The US military uses them to model actual campaigns against rival powers in which each side fights with the military forces that it realistically expects to have in the near future. The opponent is always the red team, and the US military is always the blue team, and this is how Ochmanek described what has happened in those war games for years now:

> When we fight China or Russia, blue gets its ass handed to it. We lose a lot of people. We lose a lot of equipment. We usually fail to achieve our objective of preventing aggression by the adversary....Everyone assumes based on 25 years of experience that we have a dominant military establishment—that when we go to war, we always win, we win big, and there isn't any question about this. And when

> you say to people, "not so fast," they are shocked, because they have not had this experience.[4]

The truth is even worse than Ochmanek describes. Over the past decade, in US war games against China, the United States has a nearly perfect record: we have lost almost every single time. The American people do not know this. Most members of Congress do not know this—even though they should. But in the Department of Defense, this is a well-known fact.

As McCain and I sat together at the end of the day, a pale winter twilight fading through the tall windows of his office in the Russell Building, he was clearly dejected. He slumped in his favorite antique chair and stared at the floor, his hands clasped together in front of his mouth.

"How do you think it would go?" McCain asked. "A war against China, I mean."

"Badly," I said.

"No, really, how would it actually unfold?"

What John McCain and I then proceeded to do deep into that darkening evening was imagine what would happen if the US military was called upon to fight China in the next few years. We agreed that the United States would not start the war unprovoked, but that a war could start, nonetheless, for any number of reasons. It might start with an incident at sea between Chinese and American warships that kills sailors on both sides and then quickly escalates. It could start with a Chinese attack on a US ally to which Washington feels obligated to respond. But no matter why a war might start, McCain and I saw it unfolding from there in much the same way.

Many of the US ships, submarines, fighter jets, bomber aircraft, additional munitions, and other systems that are needed to fight would not be near the war when it started but would be thousands of miles away in the United States. They would come under immediate attack once they began their multiweek mobilization across the planet. Cyberattacks would grind down the logistical movement of US forces into combat. The defenseless cargo ships and aircraft that would ferry much of that force across the Pacific would be attacked every step of the way. Satellites on which US forces depend for intelligence, communications, and global positioning would be blinded by lasers, shut down by high-energy jammers, or shot out of orbit altogether by antisatellite missiles. The command and control networks that manage the flow of critical information to US forces in combat would be broken apart and shattered by electronic attacks, cyberattacks, and missiles. Many US forces would be rendered deaf, dumb, and blind.

While these attacks were under way, America's forward bases in places like Japan and Guam would be inundated with waves of precise ballistic and cruise missiles. The few defenses those bases have would quickly be overwhelmed by the sheer volume of weapons coming at them, with many leaking through. Those bases would have no defense against China's hypersonic weapons, which can maneuver unpredictably, fly at five times the speed of sound, and strike their targets within minutes of being launched. As all of these missiles slammed into US bases, they would destroy fighter jets and other aircraft on the ground before US pilots could even get them airborne. They would crater runways, blow up operations centers and fuel storage tanks, and render those US forward bases inoperable. If any aircraft did manage to escape the Chinese missiles, it would be forced to relocate to another base in the region, which itself would come under attack. It would look like a US evacuation.

In the early days of a war with China, many of the forces located at these forward bases would not even be in the fight. Older, non-stealthy fighter jets, such as F-15s and F-16s, would not play an offensive role, because they could not survive against China's advanced fighters and surface-to-air missile systems. The same is true of the Navy's F-18s. The limited numbers of stealthy, fifth-generation fighter jets that could be brought to bear, such as F-22s and F-35s, can fly only several hundred miles on a single tank of fuel, so they would depend heavily on aerial refueling tankers to be able to reach their targets. But because those tankers are neither stealthy nor equipped with any self-defense capabilities, they would be shot down in large numbers. With those aircraft lost—which the Air Force never assumed could happen when they were developed—there would be no backups to keep America's short-range fighter jets in the fight.

A similar dynamic would play out with America's sea bases. Once the war started, US aircraft carriers in the region would immediately turn east and sail *away* from China, intent on getting more than a thousand miles away from the opponent's long-range anti-ship missiles. But from that far away, none of the aircraft on the flight deck would be capable of reaching their targets without aerial refueling, so the Navy would find itself on the horns of the same dilemma the Air Force faced: its stealthy fighter jets would be pushed so far back that they could only get to their targets with the help of non-stealthy, defenseless refueling aircraft that would be shot down in large numbers.

All the while, Chinese satellites and radars would be hunting for those aircraft carriers as well as additional carriers meant to provide reinforcement that would begin their long journey across the Pacific Ocean from wherever they were in the continental United States. If found, those ships would face large salvos of Chinese missiles, especially the DF-21 and DF-26 anti-ship ballistic missiles, better known in US defense circles as "carrier killers." The carriers and their escort ships might shoot down some of the missiles, but there would be so many that some could get through and knock the carriers out of the fight by cratering their flight decks, damaging their control towers, or destroying their aircraft before they even got airborne. It is also possible that a hit could be fatal, sending five thousand Americans and a $13 billion ship to the bottom of the ocean—all at the cost to China of around $10 million per missile.

The Marine Corps would struggle even more than the Navy but for the same reasons. Billions of dollars' worth of amphibious assault capability, built to deliver US troops onto enemy beaches as they had done for the D-Day landings in 1944 or the forced entry at Inchon at the start of the Korean War, would play no such role. No US commander would order a multi-billion-dollar amphibious ship a few miles off the coast of Chinese-defended territory to begin an assault while US aircraft carriers were steaming in the opposite direction to get out of range of China's missiles. Marine forces would instead aim to disperse around the Pacific and fight an expeditionary war, but they would lack many of the weapons and logistical forces to do so.

Many of the most effective forces the US military would rely on to do the heavy lifting—submarines, long-range bombers, and ground-launched missiles—would not be in the Pacific when the war started. They would need to get there first, which could take days or weeks. But even when they arrived, there would be fewer of these systems than were needed, a result of years of underinvestment and acquisition delays. And for the same reason, the systems that did join the fight would quickly run out of the most important weapons they need to be effective.

McCain and I paused and considered the potential scale of this disaster. Thousands of Americans lost in action. American ships sunk. Bases reduced to smoking holes in the ground. Aircraft and satellites shot out of the sky. A war that could be lost in a matter of hours or days even as the United States planned to spend weeks and months moving into position to fight.

After a long silence I spoke up. Imagine how that meeting in the Situation Room would go, I said, if a future president, whose name could well be Donald Trump, came to realize that the only available options are surrender and lose or fight and lose. The bigger question at that point would be whether that future president would even be willing to go to war at all. After all, that has been China's goal all along—as Sun Tzu counseled in *The Art of War,* to "win without fighting."

McCain looked as solemn and dispirited as I had ever seen him—not just physically frail from his illness and its treatments but also as if something larger were weighing on him. I could not myself help thinking about how much we had done over all of those years together to try to address this problem—all of the additional resources we had helped to secure for the military, all of the investments in new technologies and capabilities we had made, the many times we had tried (and often failed) to divest of old military systems to make room for new ones, all of the reforms we had shepherded through Congress to try to get better technology into the hands of America's troops faster. And yet, all of it seemed so unequal to the scale of the problem.

"I just don't understand," McCain murmured into his hands. "I remember when the Chief of Staff of the Army testified to Congress in 1980 that we had a 'hollow' force. It was shocking. It was front-page news." McCain paused. "What is happening now is just as bad," he continued. "It is actually much worse. And no one seems to care. They don't even seem to *want* to know."

McCain looked away from me and stared at the floor, and I will never forget what he said next. "Future generations of Americans are going to look back at us," he lamented, "and they're going ask how we let this happen, and why we didn't do more about it when we had the chance."

I know what you are probably thinking—there is no way this can be true. The United States spends close to three-quarters of one trillion dollars on national defense *each year.* That is more than the next eight countries spend *put together.* That money buys a lot of military capability—fighter jets, submarines, aircraft carriers, battle tanks, attack helicopters, nuclear weapons, and hundreds of thousands of incredibly well-armed people. Most of these military systems are capable of remarkable technological feats. They have enabled the United States of America, when necessary, to go anywhere, at any time, and do anything to any opponent in the world. The thought that *this* military, *our* military, could not win in the future—that just seems impossible.

But it *is* possible. And it is about to get much, much worse. The question that many Americans are right to ask is, Why? And how can we change course before it is too late?

For the past decade, I have worked within America's defense establishment—the iron triangle of the Department of Defense, Congress, and defense industry that McCain, modifying President Dwight Eisenhower, used to call the "military-industrial-congressional complex." In this time, I have come to believe there is a systemic failure in our defense establishment, a world I still inhabit. It is a failure to understand what is really happening in the construction and exercise of military power, and this failure leads us to misjudge and mismanage our defense enterprise.

All too often in defense, we think the measure of our strength is our platforms—individual vehicles and specific advanced military equipment and systems. We generate our requirements for military power in terms of platforms. We build our budgets and spend our money on the basis of platforms. We define our goals for military capability in relation to platforms. We aspire, for example, to a 355-ship Navy or a 386-squadron Air Force. We are drawn to platforms, in large part, because they are tangible. We can count them, touch them, and employ people to build them. They look good in parades. Indeed, platforms often rise to the level of defining the very identities of our military institutions and the men and women who comprise them, who see themselves as fighter jocks, ship drivers, and tankers. In short, we mistake inputs for outcomes.

Leaders too often seem to lose sight of the larger objective—the reason *why* we would want any platform in the first place. For the goal of a military should not be to buy platforms. The goal is to buy deterrence, the prevention of war. And the only way to deter wars is to be so clearly capable of winning them that no rival power ever seeks to get its way through violence.

What enables victory in war? Platforms may be useful tools, but they are not ultimately the answer. Rather, the ability to prevail in war, and thereby prevent it, comes down to one thing: the kill chain.

The kill chain is a term that nearly everyone in the US military knows but few outside the military have ever heard of. It is, at the deepest level, what militaries do and have always done throughout the history of warfare. The kill chain is a process that occurs on the battlefield or wherever militaries compete. It involves three steps: The first is gaining understanding about what is happening. The second is making a decision about what to do. And the third is taking action that creates an effect to achieve an objective. And though that effect may involve killing, more often the result is all kinds of non-violent and non-lethal actions that are essential to prevailing in war or military contests short of war. Indeed, better understanding, decisions, and actions are what enable militaries to prevent unnecessary loss of life—both their own people and innocent civilians.

Each of these steps is indispensable. Militaries cannot make good decisions or take relevant actions if they do not understand what is happening. The ability to understand and act is fruitless without the ability to make and communicate decisions. And without the ability to act, nothing else really matters. The process is also inherently sequential: acting in advance of understanding and deciding, or making decisions prior to knowing what is going on, is how mistakes get made, and for militaries, those mistakes can be fatal. When members of the US military complete that process of understanding, deciding, and acting, they refer to it as "closing the kill chain." And when they thwart the ability of a rival military to do so itself, they call that "breaking the kill chain." How fast, how often, and how effectively militaries can do both of these things is what determines whether they win or lose.

Some may find the reference to killing disturbing and indicative of something wrong with US military culture. I disagree. The kill chain is actually one idea that can make the essence of what Americans in uniform really do more intelligible and relatable. Our military can seem opaque, confusing, and incomprehensible, especially to the many Americans who have little to no meaningful contact with it. And yet, understanding, deciding, and acting is what billions of civilians do every day in their own jobs and lives. Businesses have to understand their market, decide how to compete, and then act on their plans. Sports teams must understand their competitors, decide how best to play against them, and then pull it all together on gameday. In this way, the core tasks that Americans in uniform must perform every day are no different from those of anyone else.

And yet, our military *is* fundamentally different from any other institution in America, and the kill chain also helps to explain that too. Killing is something that few members of our military are actually called upon to do. The vast majority do jobs focused on generating understanding, facilitating decisions, and implementing a multitude of different actions, most of which have nothing to do with killing. All of these tasks, however, are fundamentally focused on succeeding in a deadly business that is unlike any other in America. No one understands that unique burden and the sense of otherness it entails more viscerally or takes it more seriously than the men and women whom the rest of the nation asks to do its killing and dying. The kill chain is a helpful reminder, both for Americans in uniform and for all of us, that the stakes of our military's ability to understand, decide, and act are often life and death.

Though the challenge of understanding, deciding, and acting in warfare is timeless, kill chain is a relatively new term that is linked to the information revolution, which began in the 1980s. Prior to the information revolution, the kill chain was largely concentrated in single military platforms. For example, the process of understanding where an enemy aircraft was, deciding what to do about it, and then acting against it all occurred within one fighter jet or air defense system. Platforms collaborated, to be sure, but for the most part, an individual platform had to be its own self-contained kill chain.

The information revolution created the prospect of what became known in the 1990s as networked warfare. New technologies transformed the collection, processing, and distribution of information, making it possible to disaggregate the kill chain. One military system might facilitate understanding, another might enable decision making, and yet another might take the intended action. Instead of concentrating all of these functions in one platform, militaries could distribute them across a "battle network" of many different military systems. Kill chain, then, more accurately described the overall process and goal, because it was an actual chain of events—

information led to understanding, which led to decision making, which led to action.

The information revolution gave rise to a belief that the world was on the cusp of a "revolution in military affairs," a rare period of sweeping technological change that overturns existing military concepts and capabilities and requires a rethinking of how, with what, and by whom war is waged. A classic example of such a revolution is the emergence of automatic rifles, modern explosives, steamships, aircraft, and other industrial-era technologies that transformed warfare in the late nineteenth century in the run-up to World War I. Many believed in the late twentieth century that information technologies would lead to a similar military revolution—something like an internet of warfare waged with battle networks, and at the center of it all would be the kill chain.

The problem is that, for many years, often while preaching the language of kill chains and military revolutions, America's defense establishment never really changed its thinking. We remained focused on building and buying platforms rather than kill chains. Even at the peak of the wars in Afghanistan and Iraq, the United States was spending hundreds of billions of dollars trying to modernize our military in many of the wrong ways. We often tried to use unproven technologies to produce better versions of the same kinds of platforms that the US military had relied upon for decades. Many of these programs turned into multi-billion-dollar procurement debacles. Some produced highly capable platforms, but these platforms rarely cohere into one battle network that can share information effectively. "The main problem," one military officer put it to me last year, "is that none of my things can talk to each other."

The result is that the US military is far slower and less effective at closing the kill chain than it can and must be. The process is heavily manual, linear, undynamic, and impervious to change. Specific military systems may be able to work together to facilitate understanding, decisions, and actions for one specific purpose, but they cannot be recomposed in different ways for other unforeseen purposes. Put simply, the means by which the US military generates understanding, translates that knowledge into decisions, and then takes actions in war have not been built to adapt.

The tragic irony is that many in America's defense establishment seem to have learned the wrong lessons from this experience. They got so badly burned by their attempts to change the wrong ways that it has made them skeptical about the utter necessity of changing the right ways. Talk of a "revolution in military affairs" has been banished. *Transformation* has become a dirty word. And two decades of war in the Middle East have only exacerbated this overreaction by putting a legitimate but all-consuming focus on present conflicts at the expense of future threats.

This situation is especially dangerous because the information revolution did not end in the 2000s. It went into overdrive, propelled by commercial technology companies that usually have little to no connection to national security. Technologies such as ubiquitous sensors, "edge" computing, artificial intelligence, robotics, advanced manufacturing practices, biotechnology, new space capabilities, hypersonic propulsion, and quantum information technologies will have sweeping economic and social implications, but they will also have profound military applications that go far beyond platforms and weapons, which is traditionally how military power is conceived. What will be so consequential about these technologies, taken together, is that they will transform the entire kill chain—not just how militaries act but also the character of their understanding and decision making.

This is not science fiction. Many of these technologies exist now. Indeed, the men and women of America's military use many of them every day in their private lives. They rely upon networks of increasingly intelligent machines to buy and receive the things they need, to order rides in minutes and move around at will, to protect their homes and control many of the processes that go on inside of them, to have all of their most important data right at their fingertips, and to receive informed recommendations from machines all day about information they may need to know and things they may want to do—in short, to improve their understanding of the world around them, help them make better and faster decisions, and assist them with more relevant actions that save them time and improve their lives.

And yet, when members of our military put on their uniforms and report for duty, hardly any of this technology is available to them. Instead, they consistently have to do dangerous and important jobs with technology that might be many years behind what they use in their daily lives. This was reinforced again for me at a major Air Force conference last year, where I spoke on a panel about how new technology could help build better networks of military systems. An airman in the audience asked the panel how this would be possible when most servicemembers currently deal regularly with long network outages that leave them disconnected from email and the internet. Nearly everyone in the audience, more than one thousand people, erupted in applause.

The problems facing the US military are now taking on a fundamentally different and greater sense of urgency, and it goes beyond emerging technologies. The reason is China.

For the past three decades, the Chinese Communist Party has gone to school on the US military and its entire way of war. It has raced to catch up. From 1990 to 2017, the Chinese military budget increased by 900 percent.[5] China has devised strategies not to beat America at its own game but to play a different game—to win by denying the US military the opportunity to project power, fight in its traditional ways, and achieve its goals. China has rapidly developed arsenals of advanced weapons intended to break apart US battle networks, destroy the US military's traditional platforms, and shatter its ability to close the kill chain. This threat has progressed much further than most Americans realize.

That is not to suggest that China is ten feet tall or that the US military has no means of responding effectively to the challenges it poses. The greater concern is where China is headed. If it continues to grow wealthier and more powerful, China could in one decade have a larger gross domestic product than America[6]—and with it, an ability to generate as much, or more, military power as the United States. This would make China not just a great power but also a *peer*. Americans have not faced such a powerful competitor since the nineteenth century, and there is no living memory in the United States for how to even think about a challenge of that magnitude.

The Chinese Communist Party aims to become the dominant power in Asia and in the world, and it believes that for China to win, America must lose. We have to lose the race for advanced technology. We have to lose jobs and influence in the global economy. We have to lose partners who share our interests and values. We have to lose the ability to stand in the way of the Chinese Communist Party's desire to make more of the world safe for its model of high-tech authoritarianism. And as the balance of power continues to shift out of America's favor, the Chinese Communist Party will likely become more expansive in its ambitions, more assertive in its pursuit of them, and more capable of getting its way, no matter how much that harms Americans.

A core pillar of the Chinese Communist Party's plan is harnessing emerging technologies to "leapfrog" the United States and become the world's preeminent power. It is undertaking an unprecedented effort, backed by hundreds of billions of dollars of state investment, to become the world leader in artificial intelligence, biotechnology, robotics, and other advanced technologies. The Chinese Communist Party is already using these technologies to build the most intrusive system of mass surveillance, social control, and totalizing dictatorship the world has ever seen. And its leaders clearly view these technologies as equally indispensable in their race to develop a "world-class military" that can "fight and win wars."

China poses far more than just a military challenge to the United States, but that broader challenge has a clear military dimension, and it is putting us in a real predicament. The problem is not that America is spending too little on defense. The problem is that America is playing a losing game. Over many decades we have built our military around small numbers of large, expensive, exquisite, heavily manned, and hard-to-replace platforms that struggle to close the kill chain as one battle network. China, meanwhile, has built large numbers of multi-million-dollar weapons to find and attack America's small numbers of exponentially more expensive military platforms. For us to continue to spend hundreds of billions of dollars in the same ways, on the same things, would be the height of folly. It would be exactly what our opponents want us to do.

No one should think that we face these problems because people in America's defense establishment are somehow stupid, incompetent, or negligent. To the contrary, the vast majority of these people are hardworking, mission-oriented Americans who are doing their best to do right as they understand it. They are wrestling with a complex kill chain of their own. They are trying to understand what America's enemies will do, what different threats may materialize, and how the future will unfold. They are trying to make large, complex, and costly decisions based on this imperfect information.

And they are acting under extremely difficult circumstances in risk-averse bureaucracies that seem inclined to resist and stymie change at every turn.

The real story of the threat to America's military, which this book aims to tell, is more complicated. It is a story of how some American defense leaders in recent decades clearly saw the need for change, but the institutions in which they served failed to deliver it, especially as we grew more consumed with the wars that followed September 11, 2001. It is a story of how the Chinese Communist Party observed the US military operations of recent decades with fear and resentment and then embarked on a systematic buildup of its own high-tech military. It is a story of how the worlds of national defense and high technology in America increasingly grew apart. At a deeper level, it is also a story of how the United States was spoiled by its own dominance—a cautionary tale of how a prolonged period without real geopolitical competition bred a false sense of invincibility. In short, it is a story of how the United States got ambushed by the future.

The result of this ambush is a reckoning that America's defense establishment has postponed for a long time. National defense has been the ultimate closed system. It is governed by a dense web of perplexing laws and policies, dominated by a handful of major companies, shrouded in secrecy, and impenetrable to aspiring new entrants. Whereas emerging technologies have recently disrupted and remade major global industries, from entertainment to commerce to transportation, national defense has remained largely unaffected. This situation is no longer sustainable. The entire model of American military power now finds itself in much the same position that Barnes & Noble or Blockbuster Video did amid the rise of Amazon, Apple, and Netflix, and this circumstance is forcing a similar choice: Change or become obsolete. Adapt or be left behind. But the consequences of failure for the United States go far beyond any adverse events in the commercial economy. What is at stake is nothing less than the security of all Americans and our closest allies.

That is why I am writing this book—to try to make better sense of these highly complex military and technological changes and how to navigate them successfully. And that is why this book is about the kill chain. When it feels as if the scope and speed of change are increasing, as it now does, it is all the more important to focus on the things that do not and must not change—on the essence of things. For militaries, that is the kill chain. New threats and new technologies change how militaries understand, decide, and act, but not the enduring centrality of those tasks.

Focusing on the kill chain can help us avoid the common error of mistaking means for ends, the tools we use for the outcomes we seek, when we think about technology. We have made this mistake a lot over the past three decades, and we cannot afford to do so again. America's defense establishment has relied on certain platforms for so long that it can be easy to think that the goal is simply to acquire better versions of those platforms. The real goal, however, is a more effective kill chain—achieving better understanding, making better decisions, and taking better actions. The question is not how new technologies can improve the US military's ability to do the same things it has done for decades but rather how these technologies can enable us to do entirely different things—to build new kinds of military forces and operate them in new ways.

Focusing on the kill chain can also help us avoid other common errors: the fetishization of new technology and the temptation to believe that technology alone will save us. It will not. New technologies are important, but not as important as new thinking. And new thinking is more likely to emerge if we remain focused on the right things. In certain circumstances, new technologies will provide the best solutions to close the kill chain. In other circumstances, better solutions might result from combining new and old technologies, or from using old technologies in entirely new ways. These are important but instrumental concerns. What matters more is our ability to understand, decide, and act faster and more effectively than our competitors.

In this book, I follow one of the cardinal rules I set for myself when I worked for John McCain: Do not present problems without also having answers to recommend. I propose potential solutions in the pages that follow, and these ideas are mine alone. I am not speaking for McCain, nor can I.

I think the rise of China is the central challenge facing the United States today. This strategic competition will require the long-term mobilization of all elements of America's national power, but that broader effort is beyond the scope of this book. My focus here is on the military dimension of this competition. The United States will not succeed in this competition through military power alone, but the absence of it would certainly lead to failure.

America's top national defense priority, possibly for decades to come, should be to prevent the Chinese Communist Party from establishing a position of military dominance in the Asia-Pacific region, the center of the global economy, and eventually beyond it. This will be difficult because America's own position of military dominance, which we have come to take for granted in recent decades, is eroding, and that erosion will likely persist if China continues to emerge as an advanced military power. This is hard for Americans to hear, and harder for us to cope with, but I think it is an unavoidable reality.

I would not be writing this book, however, if I thought all hope was lost. It is not. But an effective response requires a rather comprehensive reimagining of the ends, ways, and means of American military power. It requires the US military to focus less on fighting offensively and more on fighting defensively. It requires a US force that has been built to project military power into the physical spaces of our opponents to focus instead on countering the ability of our competitors to project military power outside of their own immediate spaces. It requires a sweeping redesign of the American military: from a military built around small numbers of large, expensive, exquisite, heavily manned, and hard-to-replace platforms to a military built around large numbers of smaller, lower-cost, expendable, and highly autonomous machines. Put simply, it should be a military defined less by the strength and quantities of its platforms than by the efficacy, speed, flexibility, adaptability, and overall dynamism of its kill chains.

I wish I could say that all of this change will be easy. It will not. The problem is not lack of money, lack of technology, and certainly not lack of capable and committed people in the US government, military, and private industry. No, the real problem is a lack of imagination.

The United States has enjoyed a position of unrivaled military dominance for so long that most Americans cannot imagine a world without it. The result is that we are not moving with anything close to the level of urgency that is required to be successful. Many good Americans within our defense establishment understand the urgency and are doing their best to build a different kind of US military. But they are laboring in institutions whose political, economic, and bureaucratic incentives are aligned with continuing to build and buy the military America has. Changing these dynamics would be hard under normal circumstances, but it is exponentially harder now, when American leaders are consumed by a level of political gridlock, distraction, acrimony, and outright dysfunction in Washington that is rare in our modern history.

And yet, I know that change is possible, even within our current system in Washington, because I have seen it done before and have contributed to it myself. Over many years, I helped McCain shift billions of dollars in the military's budget away from wasting assets and toward cutting-edge capabilities. I helped him enact landmark reforms in how the Department of Defense formulates strategy, develops technology, acquires military capabilities, organizes itself, and manages its human capital. But perhaps most important, I helped McCain and his colleagues create an opening to rethink our national defense, much of which has taken shape only in the wake of his death.

I am writing this book because I believe the only way the United States can take advantage of this opportunity to reimagine our national defense is if more Americans better understand the issues related to emerging technologies and the future of warfare. I believe debate of these issues among scholars, specialists, and experts is essential, and I certainly aim to cover these complex military and technological concepts rigorously, but it is important to me that these issues are accessible to the educated public, especially those whose sons and daughters, husbands and wives, fathers and mothers could bear the burden of future war. I believe that broader understanding is the only way these important national challenges can be met.

And they must be met. The United States is hardly perfect. We possess plenty of contradictions and a healthy capacity to make mistakes, and our political life at present can seem uniquely acrimonious and demoralizing. I lived all of this intimately in my time with John McCain. But I am unequivocally rooting for the United States, because I would much rather live in a world where the values that define the future of technology and warfare—the future of the kill chain—are the values of the American people.

ONE

WHAT HAPPENED TO YODA'S REVOLUTION

In 1991, as the Cold War was coming to an end, Andrew Marshall was looking to the future. For eighteen years, he had been the director of one of the most influential offices in the Department of Defense that most Americans had never heard of: the Office of Net Assessment. Its mission, in brief, was to determine how the United States measured up against its competitors, primarily the Soviet Union, and how it could improve its strategic position over time. Marshall reported directly to the secretary of defense (he had already worked for seven of them) and few beyond the secretary himself were privy to Marshall's writings. Marshall would go on to work for seven more defense secretaries in a forty-two-year career that earned him a mythic status and led many in the Pentagon, including a few secretaries of defense, to refer to him reverentially as "Mr. Marshall." In Washington defense circles, however, he came to be called by a different name: Yoda.[1]

Most of the work that Marshall's office did focused on the past and present, but in 1991, he undertook a different project, and he turned to a bright Army colonel named Andrew Krepinevich for help. Marshall wanted to produce an assessment of the future of war, one that considered how technologies then emerging and the collapse of the Soviet Union would change international security, and what all that would mean for the United States. These questions became all the more pressing when, shortly after Marshall decided to begin this project, the United States responded to the invasion of Kuwait by going to war against Saddam Hussein's Iraq.

The US military force that was unleashed on Iraq was the culmination of years of refinement and planning to fight the Soviet Union, and the war in Iraq was fought from that Cold War playbook. When the decision to go to war was taken, the vast majority of US forces that were to do the fighting were based in the United States and had to be transported across the world. This "iron mountain"—ships, tanks, aircraft, and missiles, as well as all of the fuel, food, ammunition, and spare parts required to keep those forces fighting—was systematically moved into the Middle East and built up in massive forward bases. This went on, like a boxer telegraphing a punch, for six months. And the Iraqis could do nothing about it. They had no ability to contest the armies of Americans gathering at their doorstep. They had to watch and wait, and when the war started, it started on America's terms. US forces operated when, where, and how they wanted, thanks to their technological superiority. In a matter of weeks, they ran over Iraq's military.

Like many Americans of my generation, one of my formative experiences growing up was watching this war on television. As an eleven-year-old boy who was interested in military things, I was amazed not just to watch war unfold on the news in real time but also to witness the character of the war the US military appeared to be fighting. Stealth bombers flying invisibly into downtown Baghdad and striking targets that never even saw them coming. Smart bombs maneuvering down elevator shafts and through windows to knock out the exact buildings they intended to hit, without damaging surrounding areas. Tanks and attack aircraft storming across the desert, laying waste to the Republican Guard, and winning the ground war in one hundred hours. And unlike previous US wars, in which the death toll was measured in the tens or even hundreds of thousands, only 129 Americans were killed in combat by Iraqi forces.

The Gulf War, it seemed, represented a new way of warfare. But Marshall's report, which he sent to the secretary of defense in 1992, drew a different conclusion. He foresaw far more sweeping changes in how war would be fought. He had been influenced by then-classified writings of Soviet military planners about what they called a coming "military-technical revolution." Those Russian officers believed that new sensors and surveillance technologies would be able to identify all of the targets on the battlefield and feed that information nearly instantly to new weapons that could strike the targets more precisely and from farther away than ever. It would be the fastest, most effective kill chain in history. The Soviets called it the "reconnaissance-strike complex," and they believed the US military had demonstrated it in Iraq.

Marshall disagreed. His report concluded that "the United States did not come close to its potential to move the most useful information rapidly to those who needed it most." This assessment was affirmed by a major study of the Gulf War commissioned by the Pentagon and released the following year, in 1993. "Some of the aspects of the war that seemed most dramatic at the time," it concluded, "appear less so than they did in the immediate afterglow of one of the most one-sided campaigns in military history."[2] The US battle network "had not changed appreciably from the Vietnam era."[3] It struggled to find and hit moving targets, such as SCUD missile launchers. And of the 41,309 US airstrikes conducted during the Gulf War, most were dumb bombs, not smart ones. Admiral William Owens, who became vice chairman of the Joint Chiefs of Staff the following year, later concluded that "we conducted the campaign in Iraq essentially as Napoleon, Ulysses S. Grant, and Dwight D. Eisenhower had conducted their earlier campaigns."[4] In other words, massed brute force.

What Marshall saw coming was the emergence of what he later called a "revolution in military affairs." These were periods when major technological developments transformed the weapons and ways of war. Marshall pointed to historical examples, such as the inventions of machine guns, steam-powered ships, and airplanes. But he also stressed that new technology on its own did not enable militaries to succeed. They also had to develop new ways to employ that technology operationally and reform old institutions for new strategic purposes.

Marshall's report warned that we were likely "in the early stages of a change that could run another one or two decades." It would be a new revolution in military affairs, driven mainly by new information technologies that would enable better understanding, decision making, and action in war. Powerful militaries would compete for a different kind of strategic advantage. The goal would be less to amass traditional platforms and weapons than to create faster and more effective kill chains. This would require the United States to think differently about military power, Marshall argued, because a successful revolution in military affairs would call into question the efficacy of many of the US military's legacy platforms, such as tanks, manned aircraft, and large ships. Indeed, if an adversary were able to build these new kinds of kill chains, America's traditional ways and means of warfare, as demonstrated in the Gulf War, would not suffice in the future.

Shortly after Marshall finished his report, some of his analysts were searching for a shorthand way to describe how a powerful adversary might harness the emerging military revolution to counter America's traditional platform-centered approach. They settled on the term "anti-access and area denial" capabilities. Little did they know, all the way back in 1992, that these were exactly the kinds of weapons that China was beginning to build.

Marshall's call for change was prescient, and he later described the report as "perhaps the best-known assessment" his office ever prepared. Yet its publication did not have the intended effect. The problem was the defense secretary, the hundreds of millions of Americans his agency was charged with protecting, and most members of Congress seemed to see no need for change.

When the dust settled in Iraq, the United States was in a new world. The Soviet Union was gone, and what had disappeared with it was a certain reality of world affairs that had been in place for most of history: great-power rivalry and the prospect of great-power war. The new Russia was devastated. China was still poor and weak. The half of Europe that allied with the United States was focused on unifying with the other half that desperately wanted to. Japan was growing wealthy, but it was an ally, constrained by its pacifist constitution. There were no more "demons" or "villains," General Colin Powell quipped in 1991. "I am down to Castro and Kim Il Sung."

The old "bipolar world" was gone, but rather than giving way to the historical norm of a "multipolar world," with many great powers all competing for influence, we had entered a unique new era: a "unipolar world." The United States was more than a great power. It was more than a superpower. It was, in the words of one foreign leader, the "hyperpower."

In the absence of any real threats, US leaders were eager to downsize the Cold War military and harvest a "peace dividend." The only military operations that US forces conducted, and the kind to which their government increasingly sent them, consisted of peacekeeping, nation building, and humanitarian actions in places such as Somalia and Haiti—what Thomas Mahnken has called "wars for limited aims, fought with partial means, for marginal interests."[5] The Pentagon came up with a different name for them: "Military Operations Other Than War." To the extent that the United States thought it might have to fight an actual war, it was more or less planning to rerun its Iraq playbook against similarly inferior militaries in North Korea, Iran, or Iraq.

And yet, when it came to drawing attention to the idea of a revolution in military affairs, Marshall was winning the argument—or at least it appeared that way. Washington in the 1990s was enthralled with the dawning information age and all of its new technologies, which few in the US government, especially in the defense world, actually understood. *Revolution* and *transformation* became buzzwords in defense literature and government documents. A slew of ambitious efforts to reimagine the US military was undertaken in books, papers, and government plans with names like "Army After Next," "Network-Centric Warfare," "Joint Vision 2010," and "Lifting the Fog of War." The 1997 *Quadrennial Defense Review*, the Pentagon's premiere strategy document, made it an explicit priority for the Department of Defense to "Exploit the Revolution in Military Affairs."[6]

The call for change went by different names but largely pointed to the central idea, the "reconnaissance-strike complex," that Marshall had highlighted a decade prior. It was the idea that emerging technologies would enable militaries to build new battle networks of sensors and shooters that could close the kill chain more often and more rapidly than ever, that these battle networks would rely far more on advanced machines than on human beings, and that this new kill chain would render many traditional military systems vulnerable and obsolete.

Many in Washington were saying the right things, and they were certainly spending a lot of money on defense technology. But little was changing about how, and with what, the United States actually fought. The primary reason was the same as before: it simply did not strike a critical mass of Washington's decision-makers as necessary.

So, when the United States went to war to end the ethnic cleansing by Yugoslavian president Slobodan Milosevic in the Balkans, it did so largely in the same way it fought the Gulf War, and the strategy worked just fine. The US approach rested on the same assumptions. We fought on our timelines and terms. We fought from sanctuaries that our opponents could not reach. We were technologically superior to our adversaries. And we sustained few casualties on our way to victory.

Indeed, from the start of the air campaign in Bosnia in 1995 to the end of the seventy-eight-day air war over Kosovo in 1999, the United States lost only four military personnel, all of whom were killed in training, not in combat. Winning had less to do with any decisive transformation in how the United States built battle networks and closed kill chains, and far more to do with the fact that our opponent was just not that capable. Yet, all the same, the experience reinforced America's sense of dominance, as well as its traditional assumptions about how to fight wars.

One person who put a particularly sharp point on this argument was Admiral William Owens, who retired as the nation's second-highest-ranking military officer in 1996. Four years later, he wrote a book that blasted his former profession for "residual overconfidence" and "learning all the wrong lessons from our experience" in the Gulf War. "We used victory to validate doctrine, tactics, and weapons that had prevailed against a particularly inept foe," Owens wrote. If we did not change, he warned, we risked "a major dissolution of American military strength—and perhaps even a total collapse of our military capability—in the next ten to fifteen years as weapons and equipment financed during the Reagan-era buildup two decades ago become obsolete."

Owens's criticisms were in line with Marshall's, and as the new millennium dawned, Yoda was at it again. Events in his earlier warnings were coming to pass: the United States was not doing enough to harness the revolution in military affairs to change how it fought, while other states such as China were building their own versions of the reconnaissance-strike complex.

Marshall commissioned what was called the "Future Warfare 20XX Wargame Series," which sought to realistically simulate what future wars could be like when "the key technological and strategic trends associated with an ongoing Revolution in Military Affairs have been fully played out."[7] The war games were set at an undefined future date, 20*XX*, which was presumed to be between 2025 and 2030, and they pitted the United States against an undefined rival that was innocuously called a "large peer competitor." Everyone knew that only one country fit this description: China.

Marshall turned to two defense experts to lead this series of war games over the course of the final years of the twentieth century. One was Robert Martinage, who would later go on to serve in the Pentagon under President Barack Obama. The other was Michael Vickers, who also became a senior Pentagon official for both Presidents George W. Bush and Obama. Vickers would always be better known as the former Green Beret who once jumped out of airplanes with a miniaturized nuclear weapon lashed to his ankle and who later, as a CIA agent, helped Congressman Charlie Wilson plan America's covert war against the Soviets in Afghanistan. Vickers was depicted in the film *Charlie Wilson's War* playing chess in the park against multiple people at once.

In 2001, Martinage and Vickers submitted their final report to Marshall, and it challenged many of the Pentagon's core assumptions. The report foresaw a near future in which a competitor like China would have most of the same advanced technologies as the US military. The adversary would be able to find targets quickly and hit them with large numbers of precise weapons at very long ranges, no matter where the targets were—including in outer space. The US homeland would be struck by precision-guided missiles. US military communications and logistics networks would also come under withering attack, which would deny America the ability to fight the same way it had in Iraq and the Balkans. The only way for the US military to succeed on this future battlefield was to fully embrace the revolution in military affairs and build technologically advanced networks of forces, especially unmanned forces, that could hide more effectively and close the kill chain faster than anything the US military could do at the time. The consequence for not doing so would be an unforgiving future battlefield, where the rule would be "if you can be seen, you can be killed."[8]

This was a moment when it looked like things might finally turn. A new secretary of defense had just taken office. He openly championed the need for a revolution in military affairs. He created a new Office of Force Transformation in the Department of Defense to oversee the development of new military technologies and new ways of fighting. He signed a new defense strategy that directed the Pentagon to solve key operational problems that looked as if they had been lifted directly from Marshall's 20XX wargame series—the focus not so much on the less-capable adversaries the United States had been contending with since 1991 but rather more so on emerging great powers with increasingly capable militaries. In other words, China.

This new secretary of defense seemed to be everything Marshall had hoped for. But—in what ultimately proved unfortunate for the idea of a revolution in military affairs—that secretary's name was Donald Rumsfeld, and nine months after taking office in 2001, America was attacked.

It is often said that "9/11 changed everything," and in many respects it did. Congress opened the floodgates on military funding and a torrent of new money flowed into the Department of Defense. Rumsfeld still prioritized the revolution in military affairs, and he saw the US response to the September 11 attacks as an opportunity, as he titled a major article in 2002, for "Transforming the Military." The rapid removal of the Taliban from power in Afghanistan was a start, even though that victory was conceived less by the Pentagon than by the CIA. But for Rumsfeld, the real showcase for his idea of the revolution in military affairs—the transformational potential of small numbers of fast-moving, high-tech forces—would be Iraq.

September 11 also changed everything in that it precipitated a dramatic shift in priorities, as the Bush administration turned its full attention to waging a "global war on terrorism." The crucible of this conflict created some genuine military innovations. Through the conduct of thousands of counterterrorism missions, the Joint Special Operations Command developed new technologies and new ways to fight with them that have enabled US special operations forces to close the kill chain—turning information into understanding, understanding into decisions, and decisions into targeted actions—with a devastating speed and effectiveness that have ripped apart terrorist networks across the planet. It was this military innovation, as part of the broader counterinsurgency strategy that was introduced in 2007, that decimated Al-Qaeda in Iraq and prevented the United States at that time from actually being defeated in battle in that country.

The biggest way that 9/11 changed everything was in one of the least desirable ways: amid the shift toward counterterrorism, the emerging focus on China and the anti-access and area denial threats it was creating faded into the background. The Bush administration said that the two priorities were not mutually exclusive—but that is what they became, especially as the mistake of invading Iraq proved costlier with each passing year. In retrospect, the response to the September 11 attacks marked a strategic detour deeper into the Middle East that consumed much of the attention and imagination of the US military for nearly two decades, and largely still does.

In a deeper sense, however, 9/11 did not change everything. Rumsfeld and others claimed that the initial victories in Afghanistan and Iraq were "transformational," but in reality they represented far more continuity than change. As in 1991, US forces operated from sanctuaries on the doorsteps of our opponents, who were powerless to stop us. We controlled the timing of when the wars were fought. We could operate with near impunity. We were technologically superior. Even though the US military that went to war in Iraq in 2003 was more proficient at waging precision warfare than it had been in 1991, the extent of our conventional military dominance once again appeared exaggerated because of the conventional weaknesses of the opponents we were fighting.

The bigger problem remained what it had been since 1991: there was still no pressing need for change. Washington was as confident as ever in our dominance—overly so, in fact—and in the legacy military tools and ideas that delivered it. Even as the war in Iraq was melting down into the strategic disaster it ultimately became, much of our defense establishment seemed to filter out these troubling experiences as aberrational and not applicable to our traditional assumptions about the American way of war. We kept buying many of the same kinds of military platforms and planning to use them in many of the same ways we had since the Gulf War.

There is a pervasive belief in the US defense establishment that the reason the US military is not as prepared for the future as it should be is because its budget and bandwidth have been totally consumed by the demands of fighting terrorism after September 11, 2001. It is certainly true that the non-stop pace of operations has taken a heavy toll on the US military. It is also true that the worst attack on US soil in our history required Washington to readjust its priorities and focus on real threats over potential ones. This is part of the story—but only part.

The bigger problem has to do with a failure of imagination. For the past two decades, US leaders have spent vast sums of money on misguided ideas about military power and the deterrence of war. Too often, we have imagined that a persistent and predictable presence of US forces in numerous places around the world—rather than periodic and surprising demonstrations of new and better ways to close the kill chain—would deter US rivals from acting aggressively. The result is that we have run our military into the ground through repeated deployments of limited strategic value, and US adversaries have factored this into their plans to counter us.

At the same time, rather than building faster, more adaptable, and more effective kill chains through regular, real-world experimentation, US leaders spent eye-watering sums of money trying to transform traditional military platforms, and often failed on an epic scale. Over the past two decades, during the peaks of the wars in Iraq and Afghanistan, multiple new weapons programs were started and ultimately canceled with nothing to show for them. The Center for Strategic and International Studies stopped counting the different programs at eighteen, acknowledging that the real number is far higher. All told, the Pentagon and Congress spent more than

$59 billion on these programs during the 2000s and got no usable capability by the time the programs were canceled.[9]

This list includes the $18.1 billion that the Army spent on its Future Combat System, which was to be an array of aircraft and fighting vehicles that would redefine the future of land warfare. It includes the $3.3 billion that the Marine Corps spent on its Expeditionary Fighting Vehicle, which was to replace its legacy amphibious landing craft. And it includes a whole host of other systems, from Air Force satellites to Navy ships to new helicopters for multiple military services. They were all billed as leap-ahead, next-generation technologies. They all cost billions of dollars. They all failed. They were all canceled. And they all had the same result: nothing.

To this list of canceled programs that yielded nothing must be added a longer, costlier list of programs that were initiated or accelerated when the defense spending floodgates opened in 2001. Most remain many years behind schedule and billions of dollars over budget, and yet they continue to stagger on—two steps forward, one step back—like zombies. This list includes programs that regularly make the news, such as the F-35 Joint Strike Fighter, the Ford class aircraft carrier, the KC-46 refueling tanker, and the Littoral Combat Ship. But it also includes a vast array of satellites, radios, communications equipment, software programs, intelligence systems, and much more that most Americans have never heard of but have nonetheless financed to the tune of hundreds of billions of tax dollars.

Many of these systems, which Rumsfeld and others billed as "transformational," were not actually transformational in the way that Marshall and like-minded thinkers intended. These systems did not represent better, faster ways to close the kill chain. They were simply new versions of old things. For example, one allegedly revolutionary component of the new aircraft carriers was electromagnetic catapults to launch aircraft rather than the steam systems of the past. Similarly, the new aerial refueling tanker enabled a person in the front of the plane to remotely steer the boom that delivered the gas rather than a person in the back of the plane who looked directly at the refueling operation. Most of these technologies were not even close to mature when Rumsfeld and others ordered that they be added to new platforms, but they were added all the same. Congress went along and funded them. Industry jumped at the chance to build them. And billions of extra dollars and several extra years later, many of these systems are still having problems.

The bigger issue is that most of these allegedly information age military systems struggle to share information and communicate directly with one another to a degree that would shock most Americans. For example, the F-22 and F-35A fighter jets cannot directly share basic airborne positioning and targeting data despite the fact that they are both Air Force programs and built by the same company. They were architected with different means of processing and transmitting information that are not compatible. It is as if one speaks Greek, and the other speaks Latin. If one aircraft identifies a target, the only way it can transmit that data to the other is how it was done in the last century: by a person speaking on a radio.

Such information-sharing problems are more the rule than the exception, not just with Air Force programs but also in the Army, Navy, and Marine Corps, and certainly between them. At a time when the Department of Defense, Congress, and the defense industry all seemed to be singing from the same hymnal about the centrality of new information technologies and the importance of the US military services operating together as one "joint force," they were nonetheless pouring billions of dollars into new military systems that achieved the opposite outcome, if they worked at all. The result has been a US military version of the Tower of Babel.

Some military capabilities put into development in this two-decade period actually were revolutionary, but many ended up suffocated in their cradles. The Defense Advanced Research Projects Agency (DARPA) and other organizations did promising work on artificial intelligence and other advanced technologies, but these projects rarely moved beyond the laboratory. The semiautonomous Long Range Anti-Ship Missile struggled for adequate funding year after year. For a long time, the Air Force dallied with unmanned combat air vehicles, such as the X-45, before abandoning them.[10] The Navy developed the X-47, the first unmanned aerial vehicle to be launched and recovered from an aircraft carrier, which was so effective at dropping its tail hook and landing each time in the exact same place that it damaged the flight deck, and the Navy had to program it to land in different locations on its aircraft carriers. But the X-47's success did not stop the Navy from canceling the effort a few years later. None of these development decisions created strong incentives for traditional defense companies to prioritize next-generation technologies.

These and other promising technologies were not neglected or abandoned for lack of funding but rather because they threatened traditional ideas and interests, such as manned military aviation. Even during the wars in Iraq and Afghanistan, Washington spent plenty of money on attempts to modernize the military. But most of that money went toward new versions of old things, and that is the real reason why so many truly revolutionary efforts ended up underfunded or discarded: they threatened the big, manned legacy systems that formed the identity of the military services. Many in Washington wanted to keep buying and building them, and some of those programs were important and successful. But many were attempts to modernize that have been so slow to develop that the underlying technologies are no longer modern at all, making the effort akin to one giant costly leap into the past.

The deeper problem is that the Pentagon and Congress got military modernization backward. Rather than thinking in terms of buying new battle networks that could close the kill chain faster than ever, they thought in terms of buying incrementally better versions of the same platforms they had relied upon for decades—tanks, manned short-range aircraft, big satellites, and bigger ships. These were the very same systems that Marshall wrote in 1992 would be "progressively less central to military operations" because they would become large, vulnerable targets as US adversaries developed their own reconnaissance-strike complexes. And yet it was into these legacy systems that Washington poured billions of dollars, year after year. And because Washington focused on means more than ends, pieces more than networks, platforms more than kill chains, the US military has ended up with an array of sensors and weapons that often struggle to communicate with one another and function together—like a box of mismatching puzzle pieces that only fit together, if they ever do, through large amounts of time and human struggle.

This outcome, sadly, was less the result of a bug in the defense establishment than a feature of its business model. The military services, Congress, and defense industry mainly conceive of military power in terms of platforms. The ability of these things to share information is often an afterthought. In fact, the incentives usually cut the other way: Defense companies have profited more by building closed systems of proprietary technologies that make the military more dependent on a given company to maintain and upgrade those platforms for the decades they are in service, which is where companies make their real money. This behavior stems not from malice but a rational pursuit of self-interest in a platform-centered defense market.

And that is the tragic irony of what happened to the revolution in military affairs. The military procurement programs that the Pentagon and Congress prioritized under the banner of revolution were often the opposite of what Marshall and others had envisioned. As many of those programs became costly disasters, Washington turned against the entire idea of a military revolution.

What finally did in the revolution in military affairs was the war in Iraq. While Rumsfeld and other leaders in Washington were failing to prepare for the future, they were also failing to provide US troops fighting in Iraq and Afghanistan with the body armor, blast-resistant vehicles, and unmanned surveillance aircraft that servicemembers desperately needed and demanded to fight against low-tech insurgents. The more the war dragged on, the more *revolution* became a dirty word. The number of American causalities rose amid shortfalls of manpower and basic equipment, and the idea that technology would transform future warfare came to seem synonymous with a failure to plan for the actual wars in which US troops were fighting and dying. To the many tragedies we experienced in Iraq can be added this: the war did not showcase a revolution in military affairs, as Rumsfeld and others had originally intended, but it did end up tarnishing the very idea of one.

The counterrevolution was swift and severe. By 2008, Secretary of Defense Robert Gates spoke of a scourge that he called "Next War-itis—the propensity of much of the defense establishment to be in favor of what might be needed in a future conflict." In reality, Gates got it backward. The real affliction ailing the defense establishment was "Last War-itis," the belief that the concepts and weapons that had succeeded in the past would remain successful in the future, and the willingness of that defense establishment to spend tons of money trying to optimize the past.

The Pentagon kept planning to fight in the same ways: technologically inferior enemies, uncontested battlefields, iron mountains, slow kill chains, and little attrition in combat. The Army, Navy, Air Force, and Marine Corps kept plowing money into the same kinds of weapons that they had relied on since 1991. Congress kept goading them on, adding more money to programs than the Pentagon requested, making it harder to develop different technology. Together they went about divesting the US military of weapons they believed, based on past experience and present demands to save money, would be less necessary in the future, such as air and missile defenses, long-range precision strike weapons, electronic warfare, and attack submarines. These happened to be many of the same systems that Marshall had championed back in 1992 as *more* necessary for a future in which the United States might face a great-power opponent with its own reconnaissance-strike complex.

"Last War-itis" only intensified under President Barack Obama. His overriding belief was that America remained militarily dominant, that our traditional assumptions about how and with what we would fight were valid, and that our greatest threat was not foreign rivals with new kill chains but the misuse of our own power. The top priority was to end the wars that Bush began, cut military spending, and focus on "nation-building at home." The new president did call for a "pivot to Asia," which was code for competing more seriously with China. And his Pentagon did begin work on a new operational concept called "AirSea Battle," which sought to combine air and maritime forces to confront the kinds of military challenges that China was increasingly posing in the Western Pacific. However, shortly into his presidency, Obama embarked upon a series of decisions that would render these nascent initiatives largely hollow.

In 2011, in a bid to cut federal spending, the president and Congress enacted a plan that ultimately mandated $1 trillion in cuts to the defense budget over ten years. This set off a mad scramble in Washington and the broader defense establishment to get a piece of the shrinking pie. The result was a zero-sum fight between the needs of the future and the demands of the present (the latter of which were really the priorities and programs of the past). There was not enough money for both. Present needs had armies of powerful supporters: the military wanted them, the defense industry wanted to build them, and a lot of congressional constituents benefited from them. So, perhaps not surprisingly, it was the future that ended up without a seat when the music stopped. Legacy systems were largely prioritized over new technologies, and Washington doubled down on the same old assumptions about warfare it had been making for two decades.

When the future, once again, did not turn out as US leaders assumed it would, the trend only deepened. For all of Obama's criticisms of Bush, he had begun his presidency with much the same plan—avoid nation building, improve relations with Russia, and compete more seriously with China—and he soon found himself in many of the same places Bush did: mired in tensions with Russia, unable to focus on China, and fighting wars not only in Afghanistan and Iraq but also in Libya, Yemen, Syria, and elsewhere. Only he was trying to do it all on a reduced budget. Rather than making hard choices about what not to do or buy at present in order to be more ready for the future, the administration and Congress kept handing the US military more missions, but no more resources. The military became more enslaved than ever to the tyranny of current operations.

The deeper problem that has resulted from our experiences and choices of the past three decades is that the way the United States has built its kill chains is at risk of becoming irrelevant to the future of warfare. The connections between our military systems tend to be highly rigid, excessively manual, rather brittle, and thus slow. We have largely focused on connecting specific military systems together to generate understanding, facilitate decisions, and take actions against specific kinds of targets. But those kill chains do not easily adapt to threats that are different from those they were specifically built to address. They may be highly effective against preplanned objectives that do not change much, such as striking stationary targets in the opening days of a conflict. But our kill chains struggle to confront dynamic threats, such as moving targets, or multiple dilemmas at once.

A primary reason for this is that the means by which the US military understands the world, makes decisions, and takes actions were not built to change. To the contrary, many of those underlying technologies exist in black boxes that cannot be opened and upgraded as better technologies become available. It is as if they were frozen in time out of a mistaken belief that the US military would forever be dominant and the ability to adapt need not be a core virtue of how America constructs the means of military power.

That was hardly what Andrew Marshall envisioned when he first spoke about the coming revolution in military affairs back in 1992, and that definitely was not how the world looked when he stepped down as director of Net Assessment in 2015 after forty-two years on the job. He had spent more than half of that time warning Washington that it was unprepared for the future of warfare and how a strategic rival could harness emerging technologies to counter America's traditional ways and means of warfare. Indeed, in 1992, Marshall had warned that the US military's technological advantage "has already set in motion the hunt for countermeasures to ameliorate the problems created for defenders." That hunt had been on for more than two decades by the time Washington took serious notice. It had left itself vulnerable to an ambush by the future, and on the day that it happened, the future immediately became the present.

That day was February 27, 2014.

TWO

LITTLE GREEN MEN AND ASSASSIN'S MACE

I woke up on February 27, 2014, to a pile of panicked emails about something happening in Ukraine, a country that had been roiled for months by massive protests against Russian-backed president Viktor Yanukovych. Several days prior, the president had fled Ukraine for Russia. Friends in and out of the US government were now claiming that armed soldiers were taking over government buildings and other strategic locations in the Ukrainian territory of Crimea. Media reports soon flooded in, confirming the appearance of what were quickly dubbed "Little Green Men," masked and heavily armed military operatives who wore green uniforms without insignias. The immediate speculation was that they were Russian special forces.

To say that this caught the US government off guard is an understatement. There had been indications of Russian military movements into Crimea, but in a briefing to the Senate Foreign Relations Committee earlier that week, senior State Department officials had told a room full of senators that these military activities were not inconsistent with standard movements of Russian forces to the naval base that they controlled at Sevastopol. These US officials were not lying. They were dutifully reporting the best information that their government had.

It soon became clear how wrong Washington was. In the following days, the Little Green Men seized control of Crimea, blockaded the Ukrainian army and navy, sealed off the territory from the rest of Ukraine, and cut off the twenty-five thousand Ukrainian forces that remained on the peninsula. The new government in Kyiv soon evacuated its troops, and on March 21, Moscow annexed Crimea to the Russian Federation. It was the first time since World War II that an international border on the continent of Europe had been changed through the use of violence.

Russia's Little Green Men did not stop there. More of them soon appeared in the eastern provinces of Ukraine, running the same plays they had just run so effectively in Crimea. They incited Russian-speaking separatist groups to rise up against Ukrainian government forces. They armed and fought alongside those local paramilitaries. And as that fighting grew more difficult, the Little Green Men did more of it themselves with sophisticated Russian military systems.

This was not a Russian military that most in Washington recognized. It had highly capable weapons, such as electronic warfare systems, communications jammers, air defenses, and long-range precision rocket artillery, much of which was better than anything the US military had. And the Little Green Men used these weapons to devastating effect, waging the kind of high-speed, precision warfare that had long been the purview of the US military alone.

The stories that Ukrainian commanders recounted to me at the time were chilling. The Little Green Men could jam Ukrainian drones, causing them to fall out of the sky. They could also jam the fuses on Ukrainian warheads so they never exploded when they hit their targets, but instead landed on the ground with an inert thud. The Ukrainians talked about how the Little Green Men could detect any signal they emitted and use it to target them. Minutes after talking on the radio, their positions were wiped out by barrages of rocket artillery. Their armored vehicles were identified by unmanned spotter drones and immediately hit with special munitions that came down right on top of them, where the armor was weakest, killing everyone inside. The Ukrainians tried to dig themselves into bunkers and trenches, but the Little Green Men hit them with thermobaric warheads that sucked all of the oxygen out of those closed spaces, turning it into fuel that ignited everything and everyone inside. Entire columns of Ukrainian troops were annihilated by cluster munitions.

One story from a Ukrainian officer stuck with me. His fellow commander was known to the Little Green Men as a highly effective fighter. One day during the conflict, the man's mother received a call from someone claiming to be the Ukrainian authorities, who informed her that her son had been badly wounded in action in eastern Ukraine. She immediately did what any mother would do: she called her son's mobile phone. Little did she know that the call she had received was from Russian operatives who had gotten a hold of her son's cell phone number but knew that he rarely used the phone for operational security reasons. This Ukrainian commander, being a good son, quickly called his mother back, which enabled the Little Green Men to geolocate his position. Seconds later, while still on the phone, he was killed in a barrage of precision rocket artillery.

What emerged in Ukraine in 2014 was more than just Little Green Men; it was a battle network of sensors and shooters that closed the kill chain with remarkable speed and lethality. It was a Russian reconnaissance-strike complex. And it caught Washington off guard again the following year when it emerged in Syria, where US forces had been fighting for one year.

The United States soon found itself operating cheek-by-jowl with an advanced Russian military that posed threats that most in Washington had not thought seriously about for decades. For the first time in their entire lives, US ground forces had to wonder whether the sound of an aircraft overhead was a friend or a foe. US pilots had to contend with jammed communications, highly accurate antiaircraft missile batteries, and advanced fighter jets that routinely flew too close for comfort. Similarly, US sailors in European waters began facing off against Russian warships that aggressively closed within meters of theirs, submarines that they struggled to track, and cruise missiles that Russian vessels shot into Syria right over top of US Navy ships.

It was not long before this new Russian military was casting a darker shadow over the eastern flank of the North Atlantic Treaty Organization (NATO) alliance, which the United States had a treaty commitment to defend from attack. Russia violated the Intermediate Nuclear Forces treaty when it deployed ground-launched cruise missiles that could target Europe. It held "snap" military movements in which tens of thousands of its forces suddenly materialized in Russian military districts opposite NATO countries, leaving US commanders to wonder whether these were training exercises or the start of a real attack. Russian leaders even talked about how they would rapidly escalate a conflict with NATO states to the threat of nuclear war to deter a US intervention.

In 2016, two analysts from the RAND Corporation, David Shlapak and Michael Johnson, predicted that Russian forces could reach the outskirts of all three Baltic capitals in sixty hours and that US and NATO forces would struggle to respond effectively.[1] A troubling realization began to emerge in parts of the Department of Defense and Congress: The United States could lose a war to this new Russian military. Indeed, Russian victory could be a fait accompli.

This was not a war for which the US military would be ready. Most US combat power had been withdrawn from Europe. To pay for a decade of operations across the Middle East, while funding costly military modernization programs that were still not delivering as promised, or at all, the Pentagon and Congress had divested the US military of many seemingly unnecessary systems and weapons that, it turned out, were exactly what was necessary when the Little Green Men appeared. Washington had also invested in too many programs that would not survive the kill chains that Russian forces had shown they could close. US forces had been optimized for more than two decades to fight lesser opponents, and now they found themselves confronting a technologically advanced military competitor that seemed capable of undermining how, and with what, America fought. US generals began saying publicly that their forces were "outnumbered, outranged, and outgunned" by Russia's military.

At a deeper level, however, this was not like an earthquake that struck without warning. The tactical surprise of Little Green Men swarming in Ukraine and Syria may have been hard to foresee, but the same cannot be said of the broader surprise that their appearance represented for so many in Washington. The events of February 26, 2014, were the culmination of Russian military modernization and growing geopolitical ambition that had been under way for more than a decade—a threat that many in Washington had overlooked and at times actively downplayed. The larger significance of these events was that they awoke US leaders to the similar but far greater military challenge that had emerged in China. Russia was just the wake-up call.

The reason that Washington got Russia so wrong for so long can be traced back to high hopes—perhaps even wishful thinking—after the Cold War. The Russia that emerged from the fall of the Soviet Union was a broken country, and American leaders hoped it could be a useful ally. Its experience of the 1990s—weakness, poverty, criminality, lawlessness, and lost pride—was the polar opposite of America's: hyperpower, unipolarity, unrivaled dominance, and boundless optimism that it could remake the world in its image. American optimism extended to Russia: beginning with President George H. W. Bush and continuing through the Obama administration, a bipartisan consensus emerged that sought to shape Russia into a democratic, capitalist country and a partner of the United States. This led Washington to assist Russia with its transition from communism, security of its nuclear weapons, and entry into the G-8, the World Trade Organization, and other global institutions. But when interests diverged, as they did over Kosovo and Iraq, it was clear who was the great power and who was not. Washington got its way.

Donald Trump is the latest US president since Clinton to come into office with aspirations of building a better relationship with Russia and believing that his predecessor's failure to do so was largely the result of personal blunders and vices that he would correct. And yet, by the end of the Clinton, Bush, and Obama presidencies, US-Russia relations had reached new lows. What each president was slow to learn was that Russia was more interested in restoring the great-power status it lost in 1991 than in becoming the partner the United States hoped it would be. This was especially true once Vladimir Putin became president on New Year's Eve 1999.

As Putin deepened his hold on power in the early 2000s, he accelerated Russia's military modernization. A ruler who referred to the demise of the Soviet Union as "the greatest geopolitical catastrophe of the 20th century" set about restoring what he believed was Russia's rightful place in the global balance of military power. And as Russia's power grew, so did Putin's geopolitical ambitions.

An early trial run for both Putin's ambitions and the new military he was building came in August 2008, when he sent Russian forces into the former Soviet republic of Georgia. The result was a victory of a sort: Moscow carved off two separatist enclaves and effectively annexed them to Russia. But the operation itself was a troubling wake-up call for Putin. His new military showed itself to be inept at basic functions of warfare. It struggled to project power into territory that it had only recently controlled. The United States and NATO did not intervene to save Georgia, as they had with Kosovo, but Putin knew that if the West had stepped in, the Russian military would not have been able to stop it.

Rather than backing down, Putin doubled down. Russian planners had been studying the US military for a long time. They knew that US forces would fight a future war on Russia's periphery in much the same way they had fought past wars in Kosovo and Iraq. So, as the Obama administration was going out of its way to "reset" US relations with Russia, Putin was pouring money into the construction of an arsenal of technologically sophisticated weapons: long-range missiles and rockets, highly capable special operations forces, advanced air defenses, electronic warfare, cyber weapons, lasers to blind satellites, missiles to shoot them down, and tactical nuclear weapons. All of this military modernization had one explicit purpose: to render the United States incapable of projecting military power into Europe and defending its NATO allies, especially the many parts of Europe that Putin still believed should be part of a greater Russia.

This new Russian military performed very differently in Ukraine, and later in Syria, than it had in Georgia. Putin treated these interventions as live-fire exercises for Russia's new battle networks of unmanned aircraft, persistent sensors, precision fires, and Little Green Men. The result was a high-speed kill chain that not only decimated Ukrainian and Syrian opposition forces but also undermined many of the ways and means of the US military. These battlefields looked disturbingly similar to the wars of 20*XX* that Andrew Marshall had been encouraging people to envision nearly two decades prior. In short, Putin was building the reconnaissance-strike complex that his Soviet predecessors had only dreamt of.

It soon became clear that Russia's new approach to warfare would also extend beyond traditional battlefields to the internal affairs of rival nations. In February 2013, Russia's chief of the general staff and highest-ranking military officer General Valery Gerasimov outlined the stakes in an article that became required reading in the Pentagon after the Little Green Men appeared in Ukraine and Syria the next year. "The very 'rules of war' have changed," Gerasimov wrote. "The role of non-military means of achieving political and strategic goals has grown, and, in many cases, they have exceeded the power of force of weapons in their effectiveness."[2] Put another way, the reconnaissance-strike complex now included the ability to surveil the political and social fault lines in countries and strike directly at the heart of them through "military means of a concealed character." This included misinformation campaigns, political subversion, assassination, cyberattacks, and "active measures" using social media to tear at the fabric of diverse and democratic societies. The battlefield would now be everywhere.

This type of modern political warfare became known as the "Gerasimov doctrine." It was incubated in Georgia. It was used to devastating effect in Ukraine and then more broadly across Europe. And in 2016, it was used directly against the United States when Russia intervened in the US presidential election in an attempt to delegitimize American democracy in the eyes of its own citizens. This, too, caught Washington by surprise, and the US government failed yet again to respond in a timely way to a threat that had been building in plain sight and that had been deployed against others for more than a decade. Most of America's leaders had assumed it could not happen to us.

In 2014, while heads in Washington were still spinning over the emergence of the Little Green Men and the new Russian military, another ambush unfolded on the other side of the planet. China had long claimed as its own possession the entire South China Sea, the 1.35-million-square-mile portion of the Pacific Ocean in the center of Southeast Asia. China's government regularly told ships of the United States and other nations that they needed Beijing's permission to transit the sea. Washington had long disputed China's claim to the waters, not least because $3.4 trillion worth of global trade, much of it affecting the US economy, passes through the area each year. Tension in the South China Sea had been rising for years, but it drew only limited attention in Washington, which was largely distracted with domestic issues and the Middle East. That changed, however, right as Washington was turning its attention anew to Russia.

In 2014, China sent fleets of large dredger ships far from its shores into the South China Sea to transform shallow reefs and atolls into man-made islands. This was an assertive demonstration of Beijing's claim that the South China Sea belonged to China, especially because all of this construction occurred on territory claimed by China's neighbors. As the islands took shape, so too did other things: runways, control towers, aircraft hangars, and military-looking bases. In time, the US government began to observe and publicly call out China for arming these man-made islands with long-range radars, surface-to-air missile batteries, fighter aircraft, and other weapons. When President Obama raised these facts, Chinese president Xi Jinping denied them. The president of China stood before the world and lied through his teeth.

Just as Putin's intervention in Ukraine awakened leaders in Washington to the broader military problems Russia had created, China's island-building campaign set off an even larger awakening with respect to China. Political attention in Washington suddenly turned toward the concerns that leaders—certainly Andrew Marshall but many others as well—had been warning about for years: China, much like Russia, was developing a high-tech military that was purpose-built to confront the United States. Many in Washington also came to see the bigger problem: China had been working toward this goal for a lot longer than Russia had, and it had made vastly more progress. This was part of a bigger, longer story that Washington had largely neglected.

The United States has had a very different historical relationship with China than it has had with Russia. China and the United States had, in fact, been partners for the latter half of the Cold War, drawn together by a shared interest in balancing Soviet power. China, at that time, was emerging from the chaos of the Cultural Revolution and the Great Leap Forward. By 1978, its new leader, Deng Xiaoping, wanted to take China in a different direction, toward economic openness at home and integration into the global economy. Here, too, a bipartisan consensus about China arose in Washington.

It rested, as Asia experts Kurt Campbell and Ely Ratner have written, on "the underlying belief that U.S. power and hegemony could readily mold China to the United States' liking."[3] Washington was more confident than ever in its ability to achieve this lofty goal as it emerged triumphant from the Cold War and the Gulf War in 1991, but those events led to different conclusions in Beijing.

The Chinese military conducted a major study of the Gulf War, which included an after-action assessment in Iraq. Upon visiting Baghdad, Chinese military officials learned that Saddam Hussein had the same aging Soviet air defenses and other weapons that China did, and in some cases, Iraq's were better. The final study was briefed to the Central Military Commission over two days in Beijing, and in a unique commitment of personal time by a leader, Jiang Zemin, who then chaired the commission and who would soon ascend to the presidency, attended the entire two-day briefing. What unnerved the Chinese Communist Party was not just the stealth and precision of US forces but also their ability to achieve victory without completely annihilating the Iraqi military.[4] The United States had reached into Iraq and selectively destroyed its ability to fight—to close its own kill chains. And if the US military could do that to Iraq, it was not hard to imagine that it could do the same thing to China.

The years that followed provided further evidence, as Chinese leaders saw it, that they needed to transform their military and that their main threat was the United States. In 1996, as tensions between China and Taiwan flared, the United States sailed an aircraft carrier battle group into the Taiwan Strait, one hundred miles from China's mainland, and the Chinese military struggled to locate its exact position. Three years later, China watched again as the same US way of war that had triumphed in Iraq destroyed Serbia's ability to fight and forced Milosevic to capitulate. This time, however, it was personal, because a US airstrike had destroyed the Chinese embassy in Belgrade. Although Washington insisted it was an accident, Beijing disagreed, and the attack motivated the Chinese Communist Party to build a military that could stand up to America's.

Under what it called its 995 Plan (named for the Belgrade embassy attack in May 1999), China accelerated work to build a different kind of military. It continued to spend money on traditional military systems, such as ships and tanks, but its priority was to develop what it called "Assassin's Mace" weapons. The name refers to special weapons that were used in Chinese history to defeat more powerful adversaries. It would be like David and Goliath: the goal was not to beat the giant at its own game but to render it unable to fight by confronting its vulnerabilities.

These were exactly the kinds of weapons that Andrew Marshall's office imagined in 1992, and they came to be known by the same name from that earlier thought experiment: anti-access and area denial weapons. The idea was that, rather than fighting America directly, China would seek to attack the underlying systems and assumptions upon which the entire US military enterprise rested—to break its kill chains. China knew how Washington planned to fight, and it very methodically began building new weapons to counter America's approach to warfare.

The first problem that China sought to neutralize was the network of US military bases, primarily in Japan and Guam, where the United States had maintained forward-deployed forces since World War II. Washington assumed that these bases, like its other global bases, were sanctuaries that no adversary could reach. In the event of a war in Asia, the US military would build up its iron mountains in these forward bases, much as it had used similar bases to wage the wars in Iraq and the Balkans, and this would enable US forces to fight how, when, and where they wished. China knew that Washington assumed all of this, and it built larger and larger quantities of increasingly capable missiles, primarily medium-range and long-range ballistic missiles, to wipe out America's critical warfighting infrastructure in Asia. China's plan was to saturate US bases with more missiles than they could ever defend against.

The second major problem China sought to negate was US strike aircraft. Here, too, Beijing knew that airpower would be the primary means by which the United States would begin a war with China, and that it would be an indispensable part of any US war plan, much as it had been in Iraq and the Balkans. As a result, China developed early-warning and long-range radars to spot approaching US aircraft from as far away as possible. It also built dense and formidable networks of integrated air and missile defense systems that would aim to shoot down US planes from greater distances and high-powered jammers that would seek to destroy their ability to communicate. The goal was to make it harder and costlier for the United States to use its most effective weapons and fight in the traditional ways to which Washington had grown accustomed.

Beyond US land bases and strike aircraft, Beijing set its sights on America's other primary means of projecting military power: the aircraft carrier. The fact that aircraft carriers could move—so first had to be located—made them much tougher targets than land bases. But China knew that most US carriers were not based in Asia and would need to sail into the region from elsewhere in the event of conflict. This would be the window of time in which the Chinese military could locate and target them, before the carriers posed a threat. So, China set about building over-the-horizon radars, long-range reconnaissance satellites and aircraft, and other means of hunting America's floating airfields as they made their long journey across the Pacific Ocean.

China also developed weapons to attack US aircraft carriers and their associated "strike group." The DF-21, the world's first ever anti-ship ballistic missile, was designed to do just that—fly out more than one thousand miles, slam into a carrier, and cripple its ability to fight, if not sink it altogether. These capabilities eventually earned the DF-21 a different name: the carrier killer.

But China did not stop there. As Washington lurched from one costly military acquisition debacle to another, Beijing fielded an even more capable carrier killer missile, the DF-26, which may be able to fly twice as far as the DF-21, possibly farther, carry a larger warhead, and strike more precisely. It also fielded quiet diesel submarines and anti-ship cruise missiles that were harder to detect and defeat because they could fly low and maneuver unpredictably. All of these weapons were designed to strike right at the heart of US naval power.

An additional set of Assassin's Mace weapons focused on doing to the US military what it had done to Iraq in 1991: destroying the underlying systems that sustained the ability to wage war. In America's case, this was its communications and intelligence satellites, especially its Global Positioning System (GPS), which enabled US weapons to find their targets. It was the information networks that moved targeting data from sensors to shooters. And it was the logistics enterprise that allowed US forces to flow into theaters of operations and sustained forces in combat with food, fuel, and supplies. China built advanced aircraft, electronic attack and cyber capabilities, and more precise weapons, including antisatellite missiles, to counter the US military's ability to collect intelligence, communicate information, and command and control its forces in combat. This was all part of a broader warfighting doctrine that Chinese military officials ultimately called "systems destruction warfare." The simple idea was that the US giant could not move or fight if it were deaf, dumb, and blind.

In addition to Assassin's Mace weapons, China also accelerated development of modern tools to project military power. It began putting to sea a massive, modernized blue water navy that consisted of sophisticated guided missile frigates, submarines, and its own indigenously designed aircraft carriers. It built amphibious assault ships and landing craft that could carry Chinese marines ashore in places like Taiwan and elsewhere. And it developed long-range bombers and advanced fighter jets with air-to-air missiles that US officials publicly acknowledged could rival the best US weapons. If the premise of Assassin's Mace weapons was to deny the US military access to much of Asia, these other weapons would enable China to project its own power into the region in America's absence and exert its own military control.

China rapidly augmented its conventional military buildup with a similar buildup of nuclear weapons. After mastering miniaturization of nuclear warheads in the 1990s, China set about building a lot more of them along with ever more sophisticated means to deliver them. It developed all of the core components of what is known as a nuclear triad, the ability to launch nuclear weapons from missiles fired from land, from submarines, and from strike aircraft. Indeed, the most capable Chinese intercontinental ballistic missiles can each carry multiple nuclear warheads that can separate and independently strike numerous targets. These are capabilities that the United States has in larger quantities, but what is striking is that China's rapid nuclear buildup occurred while US policymakers were systematically neglecting and underfunding the maintenance and modernization of the US nuclear enterprise.

Some of the technologies behind China's Assassin's Mace weapons were indigenously developed, but many fell into Chinese hands as the result of a long-term and large-scale campaign of state-sponsored theft. This technology theft reportedly ranged from illegal appropriation of nuclear weapons designs in the 1990s to looting the designs of numerous other military programs of the past twenty-five years. US companies that entered joint ventures in China did so knowing that they were essentially putting their intellectual property in the hands of the Chinese state, which would use it to develop its domestic industries and its military. By plundering intellectual property and trade secrets from US military and defense contractors, China saved years of painstaking work and a small fortune in military research and development, much like a runner hitchhiking through the middle miles of a marathon.

It was often noted, for example, that China's CH-4B unmanned aircraft was a spitting image of the US Predator drone, and that its J-20 fifth-generation fighter jet looked strikingly similar to the F-35 Joint Strike Fighter. Indeed, some joked in Washington that all of the multi-billion-dollar acquisition disasters that plagued the US military were actually part of an ingenious plot to sabotage China when it tried to copy them. By 2012, General Keith Alexander, director of the National Security Agency and commander of US Cyber Command, estimated that the United States was losing a quarter of $1 trillion every year to cyber-enabled industrial espionage, much of it by China. He called it "the greatest transfer of wealth in history."[5]

Americans are inclined to scoff at China's centralized and authoritarian approach to technological innovation. However, it must be acknowledged that China has consistently beaten US projections of both the quality and the quantity of advanced military technology that it could develop and how quickly it could do so. Indeed, while Washington was talking about a revolution in military affairs and often falling short of its own goals before giving up on the idea altogether, China transformed its military in the historical blink of an eye to target the very military systems that Washington kept plowing money into year after year.

Ultimately, the strategic challenge that China poses dwarfs that of Russia. Despite all of its hostile rhetoric and saber rattling, Russia possesses orders of magnitude less conventional economic and military power than China, and that disparity grows wider every day. The larger significance of the Little Green Men's invasion of Ukraine was as a wake-up call for Washington about China, a far more formidable strategic challenge. But for Russia's intervention in Ukraine, it is unlikely that Washington would have shifted its attention to China as considerably as it did in 2014 and the years that followed.

It is tempting to think that the strategic challenge that China poses is a new development, or that it became significant only recently, starting perhaps when Xi Jinping came to power in 2012. In reality, however, China has been systematically transforming its military since 1993. This buildup has proceeded more quickly at times and more slowly at others, more overtly in some places and more covertly in others, but it has been one concerted, consistent effort.

The evidence has been there all along. As early as 1993, China declared that its military's goal would be "fighting local wars under high technology conditions." Chinese leaders openly spoke of the September 11 attacks as a "moment of strategic opportunity" that China had to seize while America was distracted. In 2007, China took the extraordinary, and quite public, step of blasting an aging weather satellite out of low-earth orbit with one of its new antisatellite missiles. And as Rush Doshi has argued, in the aftermath of the global financial crisis in 2009, President Hu Jintao further shifted China's strategy away from Deng Xiaoping's entreaty to "hide capabilities and bide time" toward a more overtly assertive role in the world.[6]

These events and a host of others could have been wake-up calls for Washington, but they were not heeded. To be sure, many US leaders and members of our defense establishment consistently attempted to shift Washington's focus toward China's rapid military transformation, including efforts in the Department of Defense to develop new ways to compete with Assassin's Mace weapons. But China never became the leading priority, and the growing avalanche of revelations about its strengthening, highly focused military force never meaningfully changed how the United States was building its military and planning to use it.

Washington's distraction from the greatest threat to American military dominance resulted partially from the long shadow of the September 11 attacks, the wars that followed, and the broader chaos of a deteriorating Middle East, which consumed the foreign policy of George W. Bush and, to a lesser degree, Barack Obama—both of whom had come into office stating their intent to focus on China. But to a larger extent, America's leaders disregarded the mounting strategic challenge that China posed because of the durability of the bipartisan consensus that the United States could shape China in its own image, primarily through increased trade and investment on favorable terms for China. This consensus extended beyond Washington and strongly influenced its actions: powerful domestic businesses and banks often pushed for security interests to take a back seat to economic ones in regard to China. Talk and action that deviated from this consensus was frequently scorned as warmongering that would antagonize Beijing.

The reason America's leaders so underestimated the threats that the Russian and Chinese militaries posed ultimately had less to do with them and more to do with us. We have been blinded by the myths we have told ourselves—that, with the end of the Cold War, the world had transcended great-power competition and conflict, that, in the words of *The 9/11 Commission Report,* transnational threats such as terrorism, not great-power rivalry, were "the defining quality of world politics."[7] We told ourselves that China and Russia wanted to be like America, and that greater exposure to US technology, business, and culture would make them into the partners we wanted. It is not that we were wrong to try to achieve these aspirations. It is that we clung far longer than we should have to beliefs increasingly at odds with the realities emerging all around us.

It was not until the ambushes of 2014, first by Russia and then by China, that things really began to change. Washington leaders were abruptly seized by what many of them began to refer to as "the reemergence of great power competition." And they started to think about how to respond. One of the main architects of that response was Robert Work, then the deputy secretary of defense and an acolyte of Andrew Marshall who had once written reports and run wargames for him. For many years, Work had grown increasingly concerned that China and Russia had been acquiring the kinds of military countermeasures that Marshall had warned about. "We are worried that our technological advantage is being eroded," Work conceded in 2015, "and it's being eroded at a relatively fast pace."[8]

As deputy secretary, Work tried to do something about it. The idea, which he called a new "offset strategy," was for the United States to harness the most cutting-edge technologies, such as artificial intelligence, to leap ahead of its strategic competitors. Although Work did not put it this way, he was trying to rehabilitate the revolution in military affairs that had fallen into disrepute in the two decades after Marshall had first articulated it. The technologies had changed radically, but the goal had not: to build high-tech, information-centered battle networks that enabled US troops to find enemy targets and take action against them faster than ever before—or as Marshall said two decades prior, "to move the most useful information rapidly to those who needed it most." In other words, it was still all about closing the kill chain.

There was one problem with Work's vision: despite the hundreds of billions of dollars that Washington had spent trying to modernize the military in the preceding decades, the technologies that Work and many others believed would be central to the US military's future were barely accessible to the Department of Defense. Although some traditional defense companies were developing some of these technologies, such as advanced missiles and directed energy weapons, many of the most consequential technologies were being developed by commercial enterprises that were not interested in providing them to the US military. How the Department of Defense ended up in this predicament is an even bigger part of this story.

THREE

A TALE OF TWO CITIES

How America got ambushed by the future is not just a story of Washington learning the wrong lessons about its dominance and failing to take seriously emerging great powers with chips on their shoulders and serious military technology ambitions. It is also a story of what President Dwight D. Eisenhower called "the military-industrial complex," the tight relationship that emerged at the dawn of the Cold War between private defense companies and the Department of Defense, as well as Congress. More broadly, it is a story of the close collaboration between the US government in Washington and the world of high technology that grew up in California in the 1950s, with Silicon Valley as its capital, and how that partnership eroded over the years, leaving the military and technology communities struggling to understand each other, do business with one another, and even feeling as if they lived in two different worlds with different values. The schism between Washington and Silicon Valley became a tale of two cities.

On April 30, 1946, as chief of staff of the Army, Eisenhower wrote a memo to the War Department describing the "general policies" of military acquisition that had won the war and that must be sustained in peacetime.[1] He called for integrating civilian technologists and industrialists with military operators, separating responsibility for buying current weapons from building future technology, and providing for "effective unified direction" of major technology development efforts. The major lesson of World War II, Eisenhower said, was that "while some of our Allies were compelled to throw up a wall of flesh and blood as their chief defense against the aggressors' onslaught, we were able to use machines and technology to save lives."[2] These were principles that Eisenhower carried into his presidency, as a new strategic contest dawned.

It is difficult to overstate the all-encompassing sense of urgency that Washington felt in the early years of the Cold War. The nation had survived the most cataclysmic war in human history only to find itself locked in a long-term strategic competition with a great power that possessed a hostile ideology and a growing nuclear arsenal. There was no "residual overconfidence" in America. If anything, the country was deeply concerned about falling behind in the development of military technology, a fear that became all the more real when the Soviet Union beat the United States into space with the launch of *Sputnik* in 1957. There was a pervasive belief that America could fail, and that failure could result in another apocalyptic war.

That threat had a way of focusing American minds. Nothing would be worse than losing the Cold War, and Eisenhower was prepared to do almost anything to prevent that from happening. In comparison, making massive bets and taking massive risks to develop ambitious new military technologies seemed totally acceptable, even essential, so that is what Eisenhower did.

The way Eisenhower saw it, Washington's primary role was to get the big things right. That started with picking the right people—not necessarily good people or nice people, but exceptional people, the kinds of people who might today be called "founders." Eisenhower believed in empowering these founders by giving them broad authority to solve clearly defined problems, providing them all of the resources and support they needed to be successful, and then holding them strictly accountable for delivering results. In short, it was a strategy of concentration—of priorities, money, effort, and, most importantly, people.

One person Eisenhower bet on was Air Force general Bernard Schriever, a German immigrant who had only recently gotten his first star when the president assigned him the mission of developing an intercontinental ballistic missile that could deliver a nuclear weapon to the other side of the planet in a matter of minutes.[3] This was not even close to being feasible in 1954 when Schriever got to work, but with Eisenhower's complete support and flush with funding, the general set up shop in an old church in California. He awarded gigantic contracts with fat margins to companies and technologists and integrated them into one military-industrial team. He scraped a space launch center out of a boggy stretch of Florida wetland called Cape Canaveral. He repeatedly blew up rocket engines and missile prototypes on the launchpad. But along the way, Eisenhower defended Schriever, got him more money when he needed it, and protected him from bureaucrats and staunch rivals, such as fellow Air Force general Curtis LeMay, who tried to kill the project at every turn, believing that missiles should never displace manned bombers (an early round of the fight over unmanned systems that continues to this day).

Eventually, Schriever and his team did the impossible: they developed the Thor, Atlas, Titan, and Minuteman missiles that could deliver nuclear weapons to precise locations on the other side of the planet in minutes. They laid the technological foundation from which America first went to space and then the moon. And they did it all, from start to finish, in just five years.

Schriever was not a singular success. To the contrary, he was one of many founders who developed seemingly impossible military technologies in the early Cold War. Edward Teller, a Hungarian refugee who had worked for Robert Oppenheimer—the founder of the atomic bomb—built the world's first hydrogen bomb. Admiral Hyman Rickover, a crotchety little man who would never make it past captain in today's Navy, overcame opposition in his own service in his quest to miniaturize a nuclear reactor that could fit into a submarine and power its operations for years deep underwater. Kelly Johnson, the hard-charging head of Lockheed's Skunkworks, developed, among many other aircraft, the SR-71 Blackbird, which flew so fast that it could outrun any missile shot at it. The SR-71 is still the fastest manned aircraft, and Johnson designed it with pencils and a slide rule. There were other defense founders as well, and in a matter of years, they built technology that sustained the United States through the Cold War.

This was how America acted when it was serious. The paramount concern was picking winners: the priorities that were more important than anything else, the people who could succeed where others could not, and the industrialists who could quickly build amazing technology that worked. Other concerns, such as fairness and efficiency, were of secondary importance. Did this approach occasionally result in waste, fraud, and abuse? Yes. But that was deemed the price of moving fast, getting things done, and staying ahead of the Soviet Union.

The military-industrial complex grew up in response to the incentives that Washington created. Everyone knew what the priorities were. Everyone knew that a ton of money was being spent on them. And nearly everyone, it seemed, wanted in on the action.

This is how Silicon Valley originated: as a start-up incubated by the Department of Defense. Margaret O'Mara, a historian and former staffer for Vice President Al Gore, has observed, "Defense contracts during and after World War II turned Silicon Valley from a somnolent landscape of fruit orchards into a hub of electronics production and innovations ranging from mainframes to microprocessors to the internet."[4] Those technologies formed the core of unprecedented new weapons. The guidance systems and onboard computers that steered US missiles into precise targets during the Gulf War in 1991 were largely thanks to Silicon Valley. Its deep military roots contributed to a culture that not only was willing to work on weapons but also embraced this work. During the 1950s and 1960s, a generation of engineers was motivated by the steady stream of challenging problems the Cold War produced, such as winning the space race. They believed their work could make them wealthier and America safer.

Things began to change, however, in the following decades, and these changes marked a dangerous shift that gradually began to constrain the rapid development of military technology. A sprawling bureaucracy materialized in the 1960s to administer and discipline the military-industrial complex. Eisenhower's more personalized approach to military acquisition and innovation, which was based on picking winners and holding them accountable, became bureaucratized amid the broader adoption of the industrial age management practices that had come into vogue in leading companies.

No one did more to further this trend than Robert McNamara, a veteran of General Motors who ran the Pentagon for much of the 1960s. Under his tenure, in the spirit of improving efficiency, new layers of oversight, analysis, and management were added, and these grew and began choking off the ability to develop breakthrough technologies quickly. For its part, Congress tied the military's hands through the budget process, making it harder to spend money in new ways or on new ideas that were not exactly what the Pentagon had "programmed" and Congress had decreed. Defense budgets were built years in advance as Washington sought to turn the preparation for war into a perfectly efficient management science.

The result was that the process of developing military technology became harder, slower, and less creative. This outcome only intensified in the early 1970s, when many engineers in Silicon Valley began growing uncomfortable working for the US government as the Vietnam War grew more divisive. By the late 1970s, innovators in the Department of Defense found themselves compelled to work around the acquisition system, rather than through it, to get good technology fast. Indeed, many of the weapons that debuted in the Gulf War, such as stealth aircraft and precision-guided munitions, were developed this way: William Perry, the Pentagon's leading technologist until 1981 and later secretary of defense, gave these programs such a high level of classification that most of the bureaucracy did not even know they existed.

Things had gotten so bad by the 1980s that a major commission was created to reform military acquisition. It was led by the founder of Hewlett-Packard and former deputy secretary of defense David Packard. One person he sought advice from was none other than General Schriever, who wrote to Packard on February 11, 1986. His verdict was devastating. Eisenhower had gotten it right, Schriever wrote, but "during the last several decades we have lost the way." The "timely fielding of qualitatively superior weapons is not being achieved," Schriever said, because now it took more than twice as long and "enormously" more money to develop them. Military procurement had become "politicized by a blizzard of legislation" and stifled by a "maze of top-down micro-management." The resulting system had become "a tapestry of confusion, delay and self-serving motivation," Schriever wrote, in which there are "more rules, requirements, documents, people, reviewers, and checkers than ever before involved non-productively in the decision-making process." In short, this was a system that would have made Schriever's achievements impossible.

What is most striking about Schriever's letter is that it could have been written today. Indeed, in the decades that followed the 1980s, the situation deteriorated further. At times intentionally, at times unintentionally, Washington further distorted the incentive structure of the military-industrial complex in ways that drove apart the worlds of national defense and advanced technology. When the Soviet threat disappeared, any sense of urgency in military acquisition went with it. The United States was the "hyperpower." The pressure to stay ahead of a strategic rival that had propelled military technological innovation during the Cold War vanished, along with much of the money that had sustained Silicon Valley's work.

In 1993, Deputy Secretary of Defense William Perry convened what became known as "The Last Supper." He called together the CEOs of major defense companies, which at the time was a lot of people. He told them each to look to their left and to their right, because in a few years' time, as defense spending came down, most of them would be gone. Perry urged the CEOs to consolidate, and that is exactly what they did. When the Cold War ended, there were 107 major defense firms. By the end of the 1990s, there were five.[5]

Whatever else can be said about Perry, he was not wrong about the money. A large and sustained reduction in federal funding for defense research and development began in the 1990s. The days of big, concentrated investments in new technologies to solve hard military problems disappeared with the Cold War. Members of Congress earmarked much of what money remained for research activities in their states and districts, and these often had more political than military value. The Pentagon, for its part, parceled development funding out in large numbers of small-dollar contracts that rarely scaled into big programs. Political leaders seemed more focused on the number of contracts they could spread around and the number of small businesses they could say they were funding. Washington seemed more interested in talking points than technology.

An arguably bigger problem was that the United States radically slowed its iterative development of new military systems. Early in the Cold War, for example, the US military acquired different new aircraft every few years. But after the Cold War, the cycle times to develop new aircraft and vehicles were frequently drawn out to more than a decade. In the case of the F-35, it has taken nearly two. More and more of America's defense spending shifted from developing new things to operating and maintaining old things. Ambitious young engineers who wanted to design new military aircraft and other systems faced the very real prospect that they might only get one or two chances of doing so in their entire careers. This created a powerful incentive for them to take their talents elsewhere, and many of them did.

None of this was a welcome development for technology companies that wanted to help solve military problems. The opportunities they found to work in the defense world increasingly amounted to little science projects and technology demos that often failed to transition into large military programs but disappeared instead into what became known as the "valley of death." These companies found it increasingly difficult to sustain themselves, attract private investment, and grow into larger companies. Not surprisingly, more and more of them were driven out, dropped out, or stayed out altogether. From 2001 to 2016, of new companies that sought to work for the US government, 40 percent were gone after three years, more than half were gone after five years, and nearly 80 percent were gone after ten years.[6]

If these troubling trends struck political leaders in Washington as particularly problematic, they did not intervene to change them. Nor did they seem overly concerned about a defense bureaucracy that took forever to field new military systems—they might have said they were, but nothing really changed, because there was no reason for it to change. The US military remained dominant either way. It was lapping the competition. Indeed, in the absence of existential threats, the far greater problem appeared to be vice in the military-industrial complex: waste, fraud, abuse, excessive corporate profits, and the exclusion of underprivileged groups.

As a result, congressional and Pentagon leaders began to optimize the defense acquisition system for another set of virtues—not speed to develop and field the best military technology, but rather transparency, fairness, social justice, ease of administration, and the endless pursuit of efficiency in planning and accounting for every dollar spent by the military or paid to industry. Picking winners came to seem archaic, even unfair. Congress passed law after law to create new processes, offices, paperwork requirements, and official homework checkers to ensure that some bad thing that happened once would never happen again. And the Pentagon added to these troubles by further tying itself into knots of red tape. All of this contributed to a downward spiral of risk aversion that made it difficult for creative people across the defense establishment and outside of it to develop and field new military capabilities.

The slowing of innovation increased the temptation in government and industry to begin programs that depended on future technological miracles. Acquisition bureaucrats, who would never operate weapons in combat, nonetheless painstakingly sought to define the exact requirements for those weapons, and the decision of who would get to build them came to rest more on which contractor could deliver "technologically acceptable" systems at the lowest price to the taxpayer rather than on technological innovation that delivered the best value to the warfighter. The Department of Defense and Congress sought to fix these existing technologies into so-called programs of record to make it easier to plan for future costs and predictably fund them each year. The downside was that these programs took on lives of their own and, once established and funded, became very difficult to dislodge with new and better capabilities.

In time, defense companies began to reflect the problems that plagued their biggest customer and its increasingly dysfunctional procurement system. As Washington focused more on the efficient production of incrementally better versions of existing weapons, as it prioritized cost accounting and ease of administration over rapid technological innovation, as it created ever more boxes to check and hoops to jump through in the unending pursuit of a fair and virtuous acquisition process, companies adapted. They chased contracts that paid them to develop only the capabilities that met the Pentagon's requirements, which frequently changed in the midst of building them, and paid them fees regardless of how long the work took or how well it was done. Defense companies spent less money on research and development and more on armies of lawyers, lobbyists, accountants, and consultants to help them comply with the Pentagon's growing acquisition bureaucracy and win more of the shrinking number of large contracts.

Many companies resented making these changes, which they felt forced into. But change they did, and they often used their influence in Washington's byzantine acquisition system to their own advantage: They underbid on contracts to develop technology and then overran on the actual costs and time to produce it. They promised things they could not deliver. And they used their political clout in the Pentagon and Congress to make it harder for new companies and new technologies to displace their programs of record. Put simply, the US government created incentives for defense companies to do the wrong things, and that is often what happened.

All of this exacerbated the defense industry's consolidation. That process may have started for economic reasons, but it steadily proceeded apace, from the lean years of the 1990s to the massive increases in spending that followed the September 11 attacks, through the drastic budgets cuts that began in 2011 to the upturn in spending in 2017. The reason was simple: year after year, Washington introduced new laws, policies, and regulations that made it harder and costlier for numerous companies to remain viable in the defense industry. For corporate leaders with fiduciary responsibilities to maximize profits for their shareholders, overcoming those costs through scale and the pricing power that consolidation could bring seemed the most rational path forward.

Indeed, this process has only accelerated. In just the past five years, Lockheed Martin bought Sikorsky, Northrop Grumman bought Orbital ATK, General Dynamics bought CSRA, SAIC bought Engility, L3 merged with Harris, and United Technologies bought Rockwell Collins only to merge with Raytheon two years later. It is estimated that seventeen thousand companies dropped out of the defense business between 2011 and 2015.[7] Many of these major moves represented gains in efficiency, but often at a cost to the effectiveness and speed of innovation. As defense companies grew larger, their creative engineers and technologists struggled to move fast and solve problems in the face of ever-expanding corporate bureaucracies. Though some leaders in Washington criticized this consolidation, it was a logical result of the incentives they had created.

National defense had become a nearly closed system that was increasingly unattractive to new companies, and the barrier to entry was seemingly insurmountable. As fewer companies were willing and able to do business with the military, and as the defense industry became more consolidated and less competitive, the Department of Defense turned to the same few companies for more of its needs. A narrowing group of voices was bound to create blind spots, and that was a main reason why Washington got the information revolution wrong. The defense establishment primarily thought (and still thinks) in terms of things—of building and buying platforms. The information revolution and the revolution in military affairs, as Marshall and others saw it, were less about things, and more about the connections between them. The revolution was about networks.

The Pentagon and Congress did not know how to buy synergy between platforms, and building connectivity is not the expertise of traditional defense companies. So, the military-industrial complex continued to do what it knew best: it built and bought things. And to the extent that Washington thought about the connections between those things, it turned to the same few companies to build them, too. It handed billions of dollars to manufacturers of vehicles, ships, and airplanes and tasked them with writing complex computer software, developing information technology, and building communications networks. Many of these efforts failed, resulting in programs that generated no usable capability or only pieces of hardware that could not connect with anything else. But the Pentagon kept spending good money after bad. Congress kept providing funds. And these zombie programs stumbled on.

The irony is that, thanks to all of those young engineers who fled the defense world or who had found it too unappealing to enter in the first place, the information revolution that the US military so desperately wanted exploded into existence in the outer world, on the other side of the country. As the US government's policies and actions pushed new technology developers away from Washington and the defense market, the opportunities of the commercial technology economy pulled them toward Silicon Valley in droves.

Starting with the internet boom in the 1990s and continuing with the new technology world that grew up after the millennium, commercial markets for the software, services, and consumer electronics that Silicon Valley was building quickly dwarfed the buying power of the Pentagon, as vast as it was. New start-ups were getting rich and growing into massive companies not because of a multi-million-dollar government market with hundreds of thousands of customers but because of a multi-billion-dollar commercial market with hundreds of millions of customers. By comparison, the profits to be made working with the Pentagon were a rounding error and not worth the excessive cost and hassle required to navigate its convoluted procurement system. Investors looking to deploy hundreds of billions of dollars simply did the math—the real returns would be made in Silicon Valley with commercial technology, not in Washington with defense technology—and so that river of money flowed to the former at the expense of latter.

Because of these basic incentives, a generation of America's best engineers, backed by a fortune of private investment, went to work building the technologies of the future, kicking the information revolution into overdrive—not with the purpose of maintaining the US military's technological edge, which by the early 2000s was already decaying amid China's accelerating military modernization, but instead to improve internet searches, optimize online advertising, and post cat videos on social media. Regardless of what these technologies were used for, they did exactly what advocates of the revolution in military affairs had been saying was needed to transform the military: they connected everything and everyone, put better information in more people's hands, and enabled them to make better, faster decisions about life and work—and cats.

What's more ironic still, the commercial technology revolution happened right as Washington's deepening involvement in foreign wars and flawed military modernization programs were turning more members of America's defense establishment against the prospect of revolutionary military change. Washington had rarely been less prepared for an ambush of new technologies, and yet, that is precisely when the information revolution went into overdrive.

FOUR

INFORMATION REVOLUTION 2.0

In 2018, I was visited in the Senate by some executives from a computing company called Nvidia, a member of the shrinking list of tech firms supportive of working in national defense. They came to discuss their partnership with the Oak Ridge National Laboratory that used their technology to build the world's fastest supercomputer, which would be able to perform 200 quadrillion operations per second (that is a 2 with 17 zeros after it). This was a remarkable and exciting achievement, but I soon became interested in something different.

Nvidia's core technology is called a graphics processing unit, which its founders created not with militaries in mind but video games. The gaming world had an insatiable appetite for ever greater computing power to run the increasingly high-resolution, high-speed, and large-scale games that developers wanted to develop and players wanted to play. Nvidia's miniaturized graphics processing units were the answer, and they enabled the explosion of modern gaming in recent years that brought to gamers' screens rich, virtual worlds filled with thousands of hyper-realistic artificial agents, all running at lifelike speeds with little to no latency.

What Nvidia soon realized was that the same computing engines that enable humans to navigate artificial worlds could also enable intelligent machines to navigate the real world. The company's graphics processing units were soon helping to lead a new revolution in artificial intelligence and machine learning with wide-ranging applications. It was by stacking Nvidia's most powerful computing cores together that Oak Ridge had built its supercomputer.

What interested me more, however, was Nvidia's role in powering self-driving vehicles. Nvidia is not the only company building mini supercomputers for autonomous vehicles, but it is certainly a leader in the field. It integrates powerful computer and graphics processing units and accelerators for artificial intelligence into a "chip" the size of a textbook that is embedded right onboard the vehicle. When equipped with well-trained machine learning algorithms, Nvidia's computers enable vehicles to make sense of the myriad events that happen every second on congested roads and perform complex, time-sensitive actions, such as maneuvering through city streets. The fact that all of this information is being processed and interpreted right where the vehicle collects it and needs it has led the kind of technologies that Nvidia and others develop to be called "edge" computing. A better description might be machine brains.

Unlike some leading American technology companies, Nvidia is open to doing business with the Department of Defense. I asked how many of its graphics processing units were operating on fielded US military systems. I was not surprised by the answer: none.

As the answer suggests, most US military systems are many years behind the state-of-the-art technology that commercial companies such as Nvidia are developing. The most capable computer onboard a US military system is the core processor in the F-35 Joint Strike Fighter, which has earned it the nickname "the flying supercomputer." The processor can perform 400 billion operations per second.[1] By comparison, the Nvidia DRIVE AGX Pegasus can conduct 320 trillion operations per second right onboard a commercial car or truck.[2] That is eight hundred times more processing power.

Compared to the rest of US military programs, when it comes to being an intelligent system, the F-35 is light-years ahead. The information that most US military machines collect is not actually processed onboard the machine itself. It is either stored on the system and then processed hours or even days later when the machine returns from its mission. Or it is streamed back to an operations center in real time, terabyte by terabyte, which places a huge burden on military communications networks. Either way, it is the job of humans, not machines, to comb through most of that data and find the relevant bits of information. In 2020, that is the full-time job of literally tens of thousands of members of the US military. When they are off-duty, they may use Nvidia's technology to play video games or even assist them on their drive home. But in uniform, they are essentially doing the same jobs that their grandparents did in World War II.

The information revolution may have started when the Department of Defense built the precursor to the modern internet, but as the void grew wider between Washington and Silicon Valley, between the defense world and the technology world, the US military simply got left behind. Meanwhile, commercial technology companies kicked the information revolution into overdrive and expanded it into nearly everything. Especially over the past two decades.

At its core, the information revolution still comprises the same basic building blocks as when that term became a buzzword in the 1990s. It is the mutually reinforcing development of sensors (which collect information), computers (which process and store information), and networks (which move information). Because improvements in one of these technologies enable, and indeed require, progress in the other two, the resulting pace of change has been exponential. This has enabled commercial technology companies—primarily, but not exclusively, in Silicon Valley—to expand the information revolution far beyond what people thought possible even two decades ago and well beyond where the US military remains today.

For instance, telecommunications companies have successfully built fast networks, transforming the standard in a matter of years from 3G to 4G, which transmits data ten times faster, and soon to 5G, which will likely be another twenty times faster than that. This ubiquitous connectivity stands in marked contrast to the level of connectivity at the Department of Defense. To be sure, the Pentagon's networks have to work in places and under conditions that commercial networks do not, but it makes those obstacles significantly harder to overcome because its networks often are built by companies whose core competency is bending metal into military platforms. As a result, military networks are like a medieval world of unpaved roads, handmade bridges, and checkpoints that inhibit more than facilitate the flow of information. The result is that most platforms and systems in the Department of Defense do not—indeed, cannot—connect to other platforms and systems, and certainly not easily, quickly, or reliably.

The same is true with sensors, which are like mechanical eyes and ears. The Department of Defense has spent billions of dollars building incrementally better versions of sensors it has used for decades—cameras, for example, that can see slightly farther and with better resolution. The result has been some exquisite sensors that are better than anything available to the public. But commercial technology has been catching up. The world is now awash in low-cost, high-quality, and increasingly miniaturized sensors—electro-optical, infrared, radar, lidar, and radio-frequency sensors that enable machines to see, as well as acoustic sensors that enable machines like Alexa or Siri to hear everything. Commercial companies have even started to develop sensors, such as synthetic aperture radars, that until now had only really existed in the US government.

This has led to some striking contrasts. Many American homes are now fitted with a network of low-cost sensors made by companies such as Nest and Ring that give one person with a mobile device real-time situational awareness of their most important places, whereas the average US military base is still defended by large numbers of people either standing watch or staring at rows of video surveillance monitors, stacked up like Hollywood Squares. Similarly, many Americans drive vehicles equipped with sensors that tell them everything that is going on around the vehicle at all times, whereas most American military vehicles do not have the same capabilities.

As sensors are proliferating on Earth, they are also blanketing it in outer space. From hundreds of miles away, commercial satellites can see objects on Earth in minute detail, and they may soon be able to identify individual faces. The number of these satellites grows by the hundreds every year. Silicon Valley is largely responsible for soon-to-be thousands of small satellites that will create an unblinking eye over the entire Earth, resulting in more real-time surveillance of the planet than ever before. Indeed, the National Geospatial-Intelligence Agency, a US intelligence agency that currently has a total of 14,500 personnel, recently estimated that it would need more than 8 million people just to analyze all of the imagery of the globe that will be generated in the next twenty years.[3]

Put simply, Silicon Valley is turning the entire world into a sensor, which has driven a never-ending demand for more computing power to store and make sense of the data that these sensors collect. Indeed, it is estimated that 90 percent of the data in the world today did not exist two years ago.[4] Commercial technology companies have solved this problem, too. Computer processing performance has nearly doubled each year since the 1990s, and with the advent of the cloud, computing power and data storage ceased to be goods that people needed to physically possess in their homes or offices. They became services, giving anyone at any time an almost unlimited ability to process and store data—anyone, that is, except for the US military. While some parts of the Department of Defense have recently adopted cloud computing, it was only in October 2019 that the department finally awarded a contract to set up an enterprise cloud, which quickly became embroiled in official procurement protests stemming from President Donald Trump's public attacks on one of the competitors, Amazon, and its founder Jeff Bezos.

Recently, however, the information revolution has moved beyond the cloud to what is called edge computing, which is the technology that Nvidia and other companies build. Ever since humans created computers, large-scale processing power was only possible through greater centralization. Tons of computer processors had to be stacked together, often in entire rooms or even buildings, to crunch and store data quickly and at scale. That is no longer the case. Computer processing has been decentralized and pushed out to the edge of the network, creating an ever-expanding network of smart systems such as vehicles, appliances, and even entire homes that collect, process, and communicate information by virtue of being connected to everything else—the so-called Internet of Things. What excited me about edge computing was how a supercomputer's worth of processing power could be spread across a vast array of military systems, creating a more resilient, secure, and operationally effective battle network.

What has made the next chapter of the information revolution possible is a fundamentally new approach to developing software that Silicon Valley pioneered long ago. It is a never-ending process of building, testing, and releasing the computer code that makes information technology work. This is why the apps and operating systems in our mobile devices are being updated around the clock.

That simply does not happen with US military systems, where hardware has always been king and software largely an afterthought. For most military systems, the schedule for hardware updates determines the schedule for software updates. After all, most of the companies building these systems are hardware companies, not software companies. This has created multiyear software development cycles that are doomed to failure. Think of how well your mobile device would work if its software and apps were updated only every several years. That's how it is for military systems. I cannot tell you the number of defense programs I came across during my time in the Senate—on which the US government had spent billions of dollars over many years—that were failing for the simple reason that their builders were not proficient in how to develop suitable, scalable, adaptable, and constantly improving software. And the result, time and time again, is that members of the US military are handed equipment whose functionality is inferior to what they use in their everyday lives.

The information revolution also created the conditions for an explosion in artificial intelligence and machine learning, which is the ability of machines to understand and learn from information independently of human commands. Artificial intelligence has been a subject of intense research interest for decades, and many of the algorithms that could enable machines to learn have been around for just as long. But two vital ingredients were missing until recently—tons of data and computing power—and the information revolution delivered them both. That made it possible to pump vast quantities of data through algorithms and train machines to perform tasks that previously only humans were capable of, such as recognizing people or specific objects in pictures.

Artificial intelligence exploded further in 2012, when a team of computer scientists led by Geoffrey Hinton demonstrated the power of "deep learning." This technology layers multiple algorithms together, at times more than a hundred, into one "neural network," where one layer in the network can pass its insights onto the next layer for further refinement. The first layer of a deep neural network might determine, for example, whether there are people in a picture, and the deeper layers could then analyze specific features to identify which individual people they are. The success of deep learning was directly related to the availability of the kinds of graphics processing units that companies such as Nvidia were developing. The processors enabled learning machines to ingest oceans of data in very short periods of time, which radically improved their ability to understand information and enabled a lot of other breakthroughs.

The private sector has improved this technology rapidly through efforts that show how software programs can learn from experience to perform specific, narrow tasks faster and more accurately than humans. Most famous of the experiments is perhaps that of Google's AlphaGo, the deep neural network that defeated the world champion of the game Go in 2016 using machine learning. Even more impressive, but less well known, is AlphaStar, also developed by Google, which beat some of the world's best players in 2018 in the real-time strategy game StarCraft II. That the software proved so successful at playing StarCraft II was particularly enlightening to me, given the ways in which that game models itself off of warfare.

In StarCraft II, a player must choose how to build military units with different powers and how to battle opponents that have built their own forces in the same way. Neither player knows what kind of military the other has built or how it will fight, and the number of moves that a player can make is—as in life but far beyond what is possible in a game of chess or even Go—mathematically enormous: 1 followed by 270 zeros. In StarCraft II, players also have to deal with high levels of uncertainty, imperfect information, long time horizons between actions and their consequences, and multiple fights unfolding on different parts of the battlefield simultaneously.

Google's AlphaStar learned to play StarCraft II by playing the equivalent of two hundred years' worth of the game in one week. It was then unleashed on a human professional player and defeated him in five straight games, despite clearly making many mistakes. AlphaStar got another two hundred years of experience before playing a higher-rated professional. Its play in the second challenge was not only free of noticeable mistakes but also remarkable in that it made decisions that human observers struggled to comprehend. AlphaStar won every game. It lost only when one of its distinctly machine advantages was taken away: its ability to see the entire battlefield at once.

There is simply nothing like this happening in the Department of Defense. Most Americans reap the benefits of machine learning every day. They use it to buy their next book, pick their next song, select the fastest route to drive home, and curate the information they consume online. They would be shocked to know how little machine learning technology, which they increasingly take for granted, has permeated the daily work of US military servicemembers, who in their jobs regularly have to perform laborious tasks manually that they turned over to machines and algorithms many years ago in their private lives.

Most of the Department of Defense is ill equipped to take advantage of machine learning in part because of how it deals with its own data. Long ago, the commercial world realized that data is the oil that fuels the digital world and the prerequisite for an intelligence revolution. Machine learning algorithms are not possible without large quantities of data, and for more than a decade technology companies have been working hard to amass stockpiles of it. Too much of the Department of Defense, on the other hand, still treats data like engine exhaust, a by-product of more important activities, which it regularly discards in large quantities. The bigger problem is that as Pentagon leaders have come to appreciate the importance of data, they have not turned to machine learning to help them make sense of it quickly and at scale, but rather have added more people to try to deal with it manually.

To be sure, the capabilities of machine intelligence should not be overstated. The achievements of artificial intelligence to date, while impressive, still fall into the category of performing narrow, repetitive tasks, albeit with growing degrees of complexity. This is a far cry from artificial general intelligence, which entails independent reasoning under conditions that are highly diverse and situationally dependent, such as a machine that is capable of doing everything that a human being can do. Such technology is still a long way off, if it is even feasible at all. The bigger concern for the Department of Defense, as artificial intelligence and machine learning continue to develop rapidly, is simply being left behind.

Beyond machine learning, and in part because of it, Silicon Valley has also expanded the frontiers of the information revolution into outer space. Since the 1950s, access to space was restricted by certain realities. Rockets could only make one-way trips, and sending them into space was like throwing away an airplane after one flight. This reality made space launch enormously inefficient, expensive, and rare. As a result, satellites were designed to last a long time, often decades, which meant they were highly complex, few in number, and extremely expensive. All of this cost and complexity traditionally limited access to space to only a few governments.

Things began to change about a decade ago, with the emergence of low-cost commercial space launch. Deep-pocketed visionaries such as Elon Musk, Jeff Bezos, Paul Allen, and Richard Branson began building new types of rockets, including reusable launch vehicles. Rockets that could make round trips into space made it possible to launch satellites far more frequently and at much lower cost, which meant that satellites themselves could be designed entirely differently.

Low-cost space launch has spawned a whole new industry in microsatellites. Rather than being large, few in number, expensive, and designed to last for decades, satellites can now be plentiful, cheaper, and designed more like mobile phones: mass-produced devices that get used for a few years and then replaced. This has enabled satellites to get much better much faster, because new technology is deployed every few years rather than every few decades. That, in turn, has created another entirely new industry: small, low-cost rockets that can launch a few microsatellites at a time. In short, in just one decade commercial technology companies in California and elsewhere overturned many core assumptions about access to space, and they are now expanding the frontiers of the information revolution beyond Earth's atmosphere.

I got a firsthand glimpse of this future a few years ago in a nondescript office park outside of Seattle. The building was easy to mistake from the outside, and I actually drove right by it at first. On the inside, however, it was all Silicon Valley start-up: open spaces, white walls, plentiful snacks, and immaculate rooms where engineers were building satellites the size of dishwashers. This was the home of SpaceX's microsatellite division and the program that it calls Starlink.

SpaceX's vision for Starlink is unsurprisingly ambitious for the company that pioneered reusable space launch vehicles: build a constellation of small satellites in low-earth orbit that can deliver high-speed communications and data networks to every part of the planet at all times. Since the dawn of the space age six decades ago, humankind has launched a total of roughly eight hundred satellites into low-earth orbit. Over the coming years, SpaceX plans to launch as many as twelve thousand and has sought government approval to launch thirty thousand more. Only a couple of these satellites were in orbit when I visited, but we used them to stream YouTube videos off of high-speed internet directly from outer space. SpaceX deployed sixty more Starlink satellites in May 2019. And they are not the only ones chasing such a future. Other companies, such as One Web and Blue Origin, are planning to send up large constellations of their own. If successful, these companies will give everyone and everything access to the internet everywhere on Earth at any time, and all that is needed to connect is a receiver the size of a pizza box.

As the information revolution has expanded into space, commercial companies have also used technology to transform manufacturing. For decades, manufacturing has most often occurred far away from the point of demand. Typical goods consist of multiple components that are made in many different places and then transported across vast logistics networks (often globally) to be assembled in a separate location into finished products that can be shipped to a consumer.

By comparison, advanced manufacturing has made it possible to produce increasingly complicated finished goods or critical components right at the point of demand, where and when users need them, with significant reductions in cost, time, labor, and logistics. One reason this is possible is the use of composite materials and methods, which enable machines to generate high-quality components that can be assembled into final products with little or no skilled human labor. In this way, manufacturing becomes akin to putting together furniture from Ikea.

A development with extraordinary military significance is additive manufacturing, which enables complex parts and even finished products to be printed in three dimensions using different kinds of materials, from low-cost plastics to carbon fiber to molten metals. This technology is already being used to print critical parts for airplanes, rockets, vehicles, and other machines. In time, people will be able to manufacture more of the things they need, right where they need them, at the push of a button, without much of the added cost, time, and human effort now required to build, assemble, ship, and warehouse manufactured goods. Indeed, it is no longer far-fetched to think that additive manufacturing will enable entire satellites to be printed in outer space, thereby eliminating the need (as well as the enormous cost, time, and risk) of launching them into orbit.

In recent years, commercial technology companies have even begun expanding the information revolution into the world of living things. The growth in computer processing and machine learning has enabled scientists to treat genomes, the building blocks of life, as just another big data problem that can now be decoded easier than ever. Indeed, since 2003, sequencing the human genome has become two hundred thousand times cheaper, and writing a genomic sequence has become more than a thousand times cheaper.[5] This has enabled, and been made possible, by the development of CRISPR and other low-cost genetic engineering technologies, which make it possible to create new genetic material and even new forms of life from scratch. An immediate application that will be of enormous interest to militaries has to do with expanding the frontiers of human performance enhancement—assessing more precisely which people are best at what kinds of cognitive and physical tasks, and then enhancing those natural abilities through individually customized medications or biotechnologies.

Another emerging frontier in the biotechnology revolution is "brain-computer interface" technology, which is exactly what it sounds like: the ability to connect the human brain to machines and control them. Elon Musk, who has founded a brain-computer interface start-up called Neuralink, has set the goal of "a full brain-machine interface where we can achieve a sort of symbiosis with [artificial intelligence]."[6] One near-term goal that Musk has defined is enabling people to type forty words per minute entirely by thinking.

Brain-computer interface can be achieved invasively, using surgical implants, but it is increasingly being done non-invasively. For example, the Johns Hopkins University Applied Physics Laboratory has demonstrated that robotic prosthetics fastened to the human body can pick up neural signals that enable amputees to control them like real appendages. The same technology that makes this possible, a combination of advanced sensors and machine learning, has also enabled humans to control other kinds of machines, such as drones—and not just one but groups of them. If the technology could be perfected, human beings could direct and oversee the operations of drones and other robotic military systems purely with their thoughts.

In recent years, companies in Silicon Valley and elsewhere have set their sights on an even more radical frontier of the information revolution—the ability to build technologies that can collect, process, and communicate information using quantum science, which concerns the bizarre properties of matter that are smaller than atoms. Subatomic particles behave differently, and more strangely, than larger forms of matter. They are capable, for example, of what is called superposition: that one subatomic particle can exist in two different physical spaces at the exact same time. Similarly, a pair of subatomic particles possess a quality called entanglement, which means they behave like mirror images of one another. Actions that affect one instantaneously affect the other, even when they are separated by large physical distances, and any attempt to manipulate either particle destroys their entanglement.

Quantum science runs contrary to the basic laws of physics, which is why Albert Einstein once called it "spooky." But it has been demonstrably proven, and there is now a big commercial push to build new kinds of quantum-based information technologies. One application is quantum sensors, which would use the property of quantum superposition to detect objects such as airplanes based on the tiny disruptions to gravity and magnetic fields that those objects cause as they move through the environment. Another application is quantum communications, which seek to use the property of entanglement to secure information. The idea is that, because two entangled particles mirror one another's behavior, and because external interference destroys the entanglement, particles could be used to build "unbreakable" encryption.

An additional application is quantum computers, which use quantum particles to encode and process information. In classical computers, information takes a binary form. It is encoded either with ones or zeros. In quantum computers, because of superposition, quantum particles can also be encoded as both ones and zeros at the same time. It sounds preposterous, but it works, and it makes it possible to encode information in three units, not just two. This makes quantum computers exponentially faster and more powerful than classic computers and enables them to solve problems that are beyond the reach of even the best supercomputers. For example, traditional encryption is based on complex math problems that classical computers would take millions of years to solve. Quantum computers could solve those problems in minutes.

It could be years, even decades, before quantum information technologies arrive, if they ever do. But commercial technology companies are spending huge amounts of money to develop these spooky systems, and if they succeed, the information revolution will enter a qualitatively new and different phase. The military implications would be as vast as they are disturbing.

Silicon Valley, it seemed, was expanding the information revolution to everything and everyone. It was transforming how billions of people around the world lived, worked, and related to machines, and it seemed everyone was benefiting to the fullest extent possible—everyone, that is, except the men and women of the United States military. Partly this was because many Silicon Valley companies were uninterested in providing their technologies to the Department of Defense, at first for economic reasons: working for the Pentagon took too long, was too frustrating, and resulted in too little revenue. But in time, economic differences hardened into ideological ones. Young founders and engineers who came of age after the Cold War had no memory of working with the US military. They had the same desire to change the world as their predecessors in Silicon Valley, and they were seized with their own version of the boundless optimism that swept America in the heady years of its "unipolar moment."

Technology seemed to be breaking down walls and bringing people together. Many in Silicon Valley began to see themselves as global citizens who had faith that people were naturally good and longed to live in peace, and that technology could make it all possible. This worldview seemed irreconcilable with that of the US military, which saw itself as the last line of defense against the immutable human capacity for evil, rapacity, and aggression. It was as if the

Department of Defense was living on Mars, and Silicon Valley was living on Venus.

The defense world in Washington often did not help matters. It was repeatedly ambushed by many of the technological disruptions flowing out of Silicon Valley and the rest of the commercial world. It missed the commercial space revolution. It missed the move to cloud computing. It missed the advent of modern software development. It missed the centrality of data. And it missed the rise of artificial intelligence and machine learning. Of course, plenty of actors in Washington had a vested interest in ensuring that the Department of Defense never capitalized on these disruptive technologies, but it is hard to overstate the degree to which Washington missed these major developments because it simply did not understand them, or even that they were possible.

But it is worse than that: When the Washington defense world eventually did become aware of these revolutions in commercial technology, it did not immediately embrace them. In many cases, it resisted them. The cases of two California-based start-ups, SpaceX and Palantir, have been illustrative in this regard, because their experiences have been nearly identical.

As Silicon Valley was mostly turning away from national defense, these two companies were the exception. The reusable rockets that SpaceX developed slashed the cost of space launch for everyone, including the US government, which had relied for many years on one defense company to launch its important and expensive national security satellites—a company whose record of successful launches was perfect, but whose price for this service was steep. Similarly, Palantir developed software that could analyze vast quantities of data and lift out important patterns and insights, which could help the US government thwart terrorist attacks by mapping their networks. The Department of Defense, in particular the US Army, had been struggling for years and spending billions of dollars to develop a comparable capability.

Both SpaceX and Palantir had cutting-edge technologies that the US military did not have, and unlike many of their Silicon Valley peers, they wanted to sell them to the Department of Defense. Both could be arrogant, pushy, and condescending at times, to be sure, but neither the Army (in Palantir's case) nor the Air Force (in SpaceX's) was eager to alter the status quo—even if it cost more (as in the Air Force's case) or did not work at all (as in the Army's case). Instead of giving up, both companies commenced multiyear fights to convince their prospective government customers to buy their technology, which was really only possible because each company had a billionaire founder who was willing and able to sustain that struggle. Even this did not ultimately work, however, and both Palantir and SpaceX had to sue their own customers to get a fair hearing. Both won and have become multi-billion-dollar companies.

Since the Cold War ended, dozens and dozens of start-ups have grown into billion-dollar businesses working in sectors such as consumer electronics, financial technology, social media, and biotechnology. In this time, three decades, Palantir and SpaceX are the only two that have achieved this so-called unicorn status in the defense sector. When people in Washington and elsewhere wonder why more engineering talent and private capital are not flowing into defense technology, the reason is not more complicated than this: three decades of data suggest that if you want to start a successful and profitable new business, defense is not the place to do it (unless you are already a billionaire). And the experience of the two start-ups that have managed to succeed at scale is not something that others in Silicon Valley or elsewhere have seemed eager to emulate.

The commercial technology world's turn away from defense was not helped by the fact that it, like the Pentagon's frequent contractors, was experiencing its own process of consolidation. Over the past fifteen years, major technology companies have bought dozens of technology start-ups: Facebook, for example, has bought Instagram, WhatsApp, and Oculus VR, among others, while Google has bought far more, including Android, YouTube, Waze, Nest, and DeepMind. As these and other "big tech" companies grew even bigger, they became larger global brands that increasingly saw the US government not as an asset but as a liability. So, as consolidation of the defense industry was resulting in fewer numbers of larger companies that were not the most capable of building cutting-edge technologies such as artificial intelligence, consolidation of the technology industry was also resulting in fewer numbers of larger companies, but this only made them more capable of building advanced technologies and less willing to provide them to the US military.

Around the same time, another unfolding drama began to drive the wedge further between Washington and Silicon Valley. The classified intelligence disclosures of Edward Snowden in 2013 hardened a belief in Silicon Valley that the US government was untrustworthy, bad for their increasingly global brands, and even opposed to their values. This contributed to a series of actions that only made matters worse: Apple's refusal to decrypt the San Bernardino shooter's iPhone in 2015 and provide the data to the FBI, Facebook's failure to control Russia's hijacking of its platform to meddle in the 2016 election, and Google's withdrawal in 2018 from Project Maven, the Pentagon program that seeks to use machine learning to process intelligence, and even the cloud computing contract. Many in Washington saw these and other actions as proof that Silicon Valley had become morally unserious and willing to elevate corporate profits above national defense, especially because many of these companies seemed more willing to work with the Chinese government than their own government. The relationship hit rock bottom.

Unfortunately, this drama unfolded at the very moment when the Department of Defense had finally begun to have its great awakening regarding advanced technology. Starting in late 2014 with Secretary of Defense Chuck Hagel and accelerating under his successor, Ash Carter, Pentagon leaders began arguing that the US military's technological advantage was eroding, and that retaining it would depend on new technologies such as artificial intelligence, autonomous systems, and advanced manufacturing.

It is difficult to overstate the tragic irony of this moment: at times intentionally, at times unintentionally, Washington had spent two decades erecting impenetrable walls between itself and Silicon Valley, walls that had the effect of preserving the status quo at all costs, only to come to the belated conclusion that the future effectiveness of the US military depended on many of those disruptive technologies that Silicon Valley was so instrumental in building but was now less eager than ever to provide to the government.

The scale of this divide has become staggering. The top five artificial intelligence companies in the United States—Amazon, Alphabet, Facebook, Microsoft, and Apple—spent a total of $70.5 billion on research and development in 2018. That is money they are investing in the future. In contrast, the top five defense companies—Lockheed Martin, Boeing, Raytheon Technologies, General Dynamics, and Northrop Grumman—spent a total of $6.2 billion. Indeed, Apple regularly sits on around $245 billion of "cash on hand," enough money to buy all five of those top US defense companies outright. The Department of Defense thus finds itself in a terrible dilemma when it comes to the core technologies that it now admits are the most important to its future effectiveness: The companies that are most able to help are not always willing to do so, whereas the companies that are willing to help are not always able to do so.

And this is perhaps the greatest irony of all: with all respect to Eisenhower, the biggest problem with the military-industrial complex is not that it became a threat to American liberty and self-government at home, as Eisenhower warned in his famous Farewell Address of 1961. The bigger problem is that over time the military-industrial complex has failed at the one job it had: to get the absolute best technology the nation has to offer into the hands of the US military so that America can stay ahead of its strategic competitors. This has not been the sole fault of the Department of Defense, Congress, or the defense industry. It has been a systemic failure that involves all three, on a bipartisan basis, and has resulted in a defense complex in Washington that has been so closed off to the wider world that it largely missed, failed to take advantage of, and even actively resisted what could be one of the most significant technological revolutions in history.

Eisenhower had directed the military-industrial complex to incredible effect, whatever misgivings he ultimately developed about it. But somewhere along the way, Washington turned against Eisenhower's risk-tolerant approach that had enabled innovators such as Schriever and others to do the impossible, and then spent decades replacing it with cumbersome, stultifying central planning processes that could not deliver great technology fast or at all. Washington sacrificed speed and effectiveness in the military-industrial complex for the hope of cost savings and efficiency, and it ended up with neither. It is as if America defeated the Soviet Union and then went about adopting the Soviets' military procurement system.

FIVE

SOMETHING WORSE THAN CHANGE

When it comes to the future of US national defense, leaders in Washington now seem focused on "the reemergence of great power competition" and are saying many of the right things about the need for military innovation and the importance of emerging technologies. The irony is that much of what is said today is strikingly similar to what has been said for the past three decades. What was old has become new again.

The difference is that the United States now finds itself in a decidedly worse position than we were in during the ebullient years after our triumphs in the Cold War and Operation Desert Storm. Our military is overly invested in large bases and expensive platforms that our rivals have spent decades building advanced weapons to attack. Many of the "transformational" procurement programs of the 1990s and 2000s are arriving so late (if at all) that the old systems they were supposed to replace are simply aging out of the force with nothing to take their place. What remains is a smaller, older force that has been so strained by years of operations overseas that it is still many years away from fully recovering. Meanwhile, one of the most significant technological revolutions in modern times, the dawning of the information age, has done too little to benefit the military.

The wide margin of error that America once enjoyed in the world is gone, and the political situation in Washington is, to put it mildly, chaotic, gridlocked, and dysfunctional. Some of this was unavoidable, but much of it was not. So, why did it happen?

It did not happen because people in the Department of Defense, Congress, and private industry acted maliciously, unpatriotically, or foolishly. The vast majority of these people worked hard to do the right things under incredibly difficult circumstances.

It did not happen because of intelligence failures. Plenty of people—not just Andrew Marshall but many others as well—saw these problems unfolding at the time, and plenty of information was available to convince those who did not.

It did not happen for lack of money. Washington has spent trillions of dollars on defense since 1991, but too often it was spent on the wrong military programs and foreign policies, and the resulting problems were exacerbated by an unwillingness of defense leaders to make hard choices about what military systems to stop buying and what military missions to stop doing.

It did not happen for lack of technology. The means to build a different and better US military have been consistently available and never more abundant than now, but too much money has been spent on old or unproven technologies in the pursuit of outdated or misguided conceptions of military power.

Nor did it happen entirely for lack of attention. It is certainly true that the September 11 attacks compelled leaders in Washington to prioritize the unique demands of counterterrorism, and they were right to do so. But the burdens of two decades of conflict have not fallen equally on the entire military, and through it all, the majority of US defense spending has gone to things other than the wars we were fighting. Ultimately, the exigencies of current operations are not the sole reason why the United States seemed so caught off guard by "the reemergence of great power competition" and so unprepared to capitalize fully on the emerging technological revolution. After all, managing competing priorities and determining their appropriate rank order is the essence of strategy.

To a large extent, the reason the United States has been so badly ambushed by the future is because the main problem we are struggling to address is incredibly difficult. Can militaries innovate and change in the absence of war? Indeed, this is the core question as the United States looks to the future of warfare.

Many of the ways that the US military *has* innovated and changed in recent years have only happened *because* it has been at war (albeit a very particular kind of war). That is why US special operations forces have devised new ways and means of combating terrorist networks, and that is also why the US Army and Marine Corps have become more proficient in counterinsurgency warfare. Change happened because there was a wartime demand for it, as well as clear consequences for failing to innovate and change. The problem is that many of the innovations the US military has developed to fight terrorist groups will likely be of limited utility for the challenge it now faces: great-power competitors with technologically advanced militaries and the prospect of large-scale, conventional conflict and strategic competition against rival states.

Militaries are unlike civilian institutions in many ways, but a primary difference is that they lack routine sources of real-world feedback on their performance. Sports teams play games that they either win or lose. Businesses have the market: if customers are not buying what they are selling, that is a good indication they need to change. None of this exists for militaries. They try to compensate with analysis, exercises, war games, and other forms of self-assessment. They experiment with new technology and new ways to use it, which is absolutely essential. But there is only so much an institution can learn about itself short of relevant, real-world performance. Indeed, the main thing that militaries exist to do—fight wars—rarely happens, and the better they are at deterring war, the less likely they are to have to fight one. That is a good thing, of course, but it makes it harder for militaries to know whether they are truly ready for the future.

Military innovation and adaptation are made more difficult because the nature of any bureaucracy is to resist change, not promote it. Military bureaucracy and culture are especially conservative, and not without reason. The wrong kind of change can cost lives. At its most extreme, however, this rigidity leads to what Norman Dixon famously called "the psychology of military incompetence," which includes "clinging to outworn tradition," a "failure to use or tendency to misuse available technology," a "tendency to reject or ignore information which is unpalatable or which conflicts with preconceptions," and a "tendency to underestimate the enemy and overestimate the capabilities of one's own side."[1]

It is extremely difficult for militaries to innovate and change in the absence of war, but it is not impossible. One example is the US Navy's struggle to exploit the full potential of aircraft carriers in the 1920s and 1930s. Coming out of World War I, the battleship was the center of naval power. The United States operated aircraft carriers during the war, but largely as auxiliaries for other naval forces. Carrier aircraft were mainly confined to serving as scouts for battleships, which would slug it out with their big guns against other battleships for control of the seas.

During the interwar years, an insurgency within the naval aviation community pushed for a more expansive role for aircraft carriers. This was led by Admiral William Moffett, who was, ironically, not an aviator himself but a battleship captain. Moffett and his fellow insurrectionists were focused on the looming threat of Imperial Japan and believed that if war were to come, the US Navy would have to push beyond the range of land-based air support to project power across the Pacific Ocean and destroy the Japanese fleet at sea. Aircraft carriers would be essential. Moffett was appointed the head of the Navy's newly established Bureau of Aeronautics in 1921, and according to one of his contemporaries, he "tackled the subject with almost fantastical zeal."[2]

Moffett made full use of analysis and war games to build his case for the revolutionary potential of aircraft carriers, but he went far beyond that. He devoted large portions of his force to experimenting at sea, enabling sailors to develop new operational concepts and tactics for how to fight with aircraft carriers. Moffett further experimented with different kinds of aircraft to perform more combat roles, and he invested heavily in those new technologies. Perhaps most importantly, Moffett fought his own leadership and peers within the Navy to promote aviators into jobs that had never been open to them, thereby seeding the bureaucracy with insurgents who would use their newfound power to drive internal change to the benefit of the kind of warfare he envisioned would become central to any victory—naval aviation.

Moffett ran the bureau for twelve years, providing continuity and sustained leadership through a heady period of change, but he could not have succeeded alone. He cultivated strong champions for his cause among powerful civilian leaders, such as President Herbert Hoover and Congressman Carl Vinson, the long-serving chairman of the House Naval Affairs Committee. Indeed, when the chief of naval operations tried to block Moffett's reappointment to a third term, Hoover himself overruled him. Moffett died eight years before the Japanese attack on Pearl Harbor, but the Navy entered that war having already done much of the hard work of adopting a revolutionary new technology, even as World War II would usher in further sweeping changes that ultimately saw the aircraft carrier replace the battleship as the centerpiece of the fleet.

Another example of military innovation in the absence of war was the development of the Assault Breaker initiative. Early in the Cold War, Washington knew that it could not marshal the sheer numbers of conventional forces necessary to stop a Soviet invasion of NATO countries, so it planned to use tactical nuclear weapons to defeat a westward onslaught by the Red Army. By the 1970s, many NATO countries had become rather unenthralled with the prospect of a US-led nuclear war in Europe to save them from Soviet domination. The US military needed a new solution. The problem was not just how to blunt a Red Army invasion but how to close the kill chain against the vast waves of reinforcements that Moscow would pour into Europe. This had to be done before these "follow-on forces" could make it to the front, or else it would be too late.

The prospect of losing the ability to deter conventional war in Europe was the impetus for a rapid technological development effort led by then secretary of defense Harold Brown, a child prodigy who graduated high school at fifteen, college at seventeen, and completed his PhD in physics at twenty-one. Leaning heavily on fellow innovators within the Air Force, the Army, and DARPA, Brown developed the makings of an entirely new kill chain—one that could look deep and shoot deep at Soviet reinforcements during a potential invasion. This included new intelligence-gathering aircraft to identify Soviet forces moving on the ground; new communications relays to pass that targeting data to weapons; and stealth aircraft and longer-range, precision-guided munitions to penetrate deep into Soviet rear areas and attack their follow-on forces. Integrated together, this is what became Assault Breaker.

It will never be known whether Assault Breaker would have worked as planned, because it was never employed during the Cold War. What is known, however, is that it deeply unnerved Soviet military planners, sowed doubt in their minds about whether they could win a war against NATO in Europe, and thereby served its most important purpose by helping to restore deterrence and prevent conflict. It was these core capabilities, first developed as part of Assault Breaker, that made their debut on the battlefield during the Gulf War of 1991, leading both the Soviet Union and Andrew Marshall to think that we were on the cusp of a new revolution in military affairs.

Why did these and other instances of military innovation in peacetime succeed? A few main reasons stand out. For starters, real change requires the definition of clear threats. Militaries need to know with as much specificity as possible what operational problems they must solve through the development of new capabilities and new ways of fighting. It is not enough for militaries to know that they must close the kill chain. They must also know against which specific threats, in which specific geographic locations, and at which specific scales and speeds they must act. These questions are often best framed by leaders at the top, but the best answers often come from the bottom up, when the lower ranks are empowered and given clear guidance to devise new ideas and try new things.

Similarly, the definition of specific operational problems is also necessary to guide and focus the development of new technologies. General Schriever succeeded, in large part, because he knew clearly which problem he needed to solve: deliver a nuclear weapon to the other side of the planet in a matter of minutes. That is also why Assault Breaker succeeded but a program such as the Army's Future Combat System did not: It became a theory of everything for everyone and eventually collapsed under the weight of the many divergent requirements it was directed to meet.

Real change in peacetime also requires extraordinary leadership, on both the civilian and the military sides. Civilian leadership alone cannot force military bureaucracies to change if they are reluctant or resistant. Donald Rumsfeld learned this the hard way, mainly because his habit of belittling senior officers earned him such disdain in the military bureaucracy. Civilians require the partnership of what Barry Posen has called "military mavericks,"[3] visionary leaders who are determined to use their unique expertise and legitimacy to change their own institutions. These mavericks rarely get very far on their own, however. They require committed civilian champions, especially in Congress, who provide money and moral support, remove obstacles from their way, and defend them from their opponents in the bureaucracy—which could include, as it did at least in Moffett's case, preventing their own institutions from firing them.

It is only when civilian leaders and military mavericks are aligned in favor of disrupting the status quo that real innovation becomes possible in the absence of war. That kind of alignment is why it worked with Moffett, Vinson, and Hoover—or, for that matter, for Eisenhower and Schriever—and it is something that has been unfortunately rare in the United States in recent decades.

Another critical point is that military innovation is never about technology alone or even primarily. What is always more important is what militaries use technology to do—how they use it to build new kinds of capabilities, operate in new ways, and organize themselves differently to take full advantage of their new ways of fighting. Much of the problem with the revolution in military affairs during the 1990s and 2000s is that it devolved into technology fetishism, as if the mere acquisition of new capabilities would by itself transform the US military. This is arguably an even bigger risk today, when emerging technologies such as artificial intelligence are often treated as magic condiments to be spread atop existing military systems. In reality, true military innovation is less about technology than about operational and organizational transformation.

Indeed, history is replete with examples of military rivals that had the same technologies, and what set them apart is how they used them and organized themselves differently. The archetypal case is that of France and Germany in the 1930s. Both militaries had tanks, radios, and airplanes. But whereas the French chose to employ those technologies as part of their effort to build better versions of the fortifications they had relied upon in World War I, Germany combined those capabilities into a new concept called *blitzkrieg,* which enabled the German army to maneuver rapidly through France's defensive positions, capturing Paris in roughly one month in 1940.

The kind of operational and organizational innovation that leads to real military change in the absence of war is rarely something that can be accomplished in the abstract. It requires constant, real-world experimentation. That means getting new technologies and capabilities into the hands of military operators and giving them the space to try to do new things with them, learn from their failures and mistakes, go back to the drawing board, and return with new concepts to test. Not only is this the best way for militaries to learn how to operate differently, it is also the most compelling way for innovators to convince holdouts, fence-sitters, and spoilers to buy in to their ideas for change. Adopting new capabilities and concepts inevitably means divesting of old ones, and most people can only be convinced to give up what they have when they see with their own eyes that the new ideas actually work better. Even then, they may still cling to the status quo and resist change.

This has been one of the biggest failings of the US defense establishment over the past few decades: we stopped doing meaningful experimentation, the kind that Moffett or Schriever would recognize. What we did too often instead was allow large bureaucratic committees to try to define in the abstract the exact requirements that new military capabilities should possess, rather than working out those requirements more iteratively by enabling military operators to experiment with new technologies. Not surprisingly, this is how we ended up with many of the procurement disasters that continue to plague our military to this day.

The US military's ability to conduct real-world, joint force experimentation was actually one of the first things to go as a result of the recent pressures stemming from increasing military operations and decreasing military budgets. There used to be an entire four-star command devoted to military experimentation: Joint Forces Command. We should not make too much of the work it actually did, but it was better than nothing, which is largely what the US military was left with in 2011, when the command was sacrificed on the altar of cost savings.

There is one reason beyond all of these, however, that is most important. It is the one non-negotiable factor without which military innovation cannot succeed in the absence of war. Put simply, militaries and their civilian leaders must believe there is something worse than change. They must believe that change is the lesser evil, and that a failure to change will realistically produce catastrophic, near-term consequences, such as the loss of a major war.

Ultimately, this is the deeper reason why America has been so badly ambushed by the future. For far too long, we have not truly believed there is something worse than change. We simply could not imagine it. We became increasingly spoiled by our own dominance and unmoored from reality, a malady that afflicted Washington and Silicon Valley alike. Americans failed to appreciate, in a real and visceral way, that the world we inhabited after the Cold War, the world of our hyperpower and unipolarity, was actually one of the most anomalous periods in world history, and sooner or later it would end. Believing we faced no meaningful threats, we felt little need to innovate or deviate from our path and little consequence for failing to do so.

This lack of urgency gradually dulled our competitive edge until it wore down into an all-consuming complacency about America's place in the world. We convinced ourselves that the period of peace and prosperity we were enjoying was uniquely the result of our virtues, our values, and our power, and that all of it would last forever. We filtered out evidence to the contrary, even when it was staring us in the face and ringing in our ears, choosing instead to clutch our preconceived notions ever tighter. Nothing, not even the September 11 attacks and all of the mistakes and heartache that followed, fundamentally jarred us out of our delusions, and over time, our arrogance begat ignorance, leading gradually to a contraction of strategic imagination and a profound forgetfulness about the persistence of tragedy in human affairs. Ultimately, this is why we failed to do so many of the things that we said were necessary.

"There are common causes for military disasters," Admiral William Owens wrote in 2000, shortly after retiring as America's second-highest-ranking military officer, "and at the heart of them lie dangerous smugness, institutional constraints on innovation, and the tendency to avoid questioning conventional wisdom." The result, Owens went on to say, is that "the side that is the most smug, the most convinced that its interpretation of the past is the best guide for the future, often turns out to be the loser in the next war."[4] These words of caution are even more salient today than twenty years ago, because the main question confronting America's defense establishment, its political leadership, and the American people is whether things are different now—whether we actually believe there is something worse than change.

I, for one, believe there is something worse than change. And the reason has everything to do with the Chinese Communist Party. This is not to say that Russia under Vladimir Putin is not a threat. To the contrary, Putin has made it abundantly clear, both in word and deed, that he means to dominate Russia's neighbors through any means necessary, weaken and fracture the NATO alliance, and meddle in the internal politics of the United States and European countries to undermine citizens' confidence in the legitimacy of their own democratic systems.

The challenge that Russia poses, however, pales in comparison to that of China. Russia will not get stronger over time. It will get weaker, which could actually make Putin more of a short-term danger. But whereas Putin has rather clear intentions but a limited ability to generate power, the Chinese Communist Party has rather unclear intentions but an ability to generate more comprehensive national power than any competitor the United States has ever faced.

The Cold War is often called a contest between great powers, and that is true enough, but the Soviet Union was never a peer of the United States. At the apex of its power, the Soviet Union's GDP was only about 40 percent that of the United States.[5] It was largely isolated from the broader international economy and lacked its own domestic base of technological innovation. The Soviet Union was powerful, but it was never America's peer.

China is becoming America's peer, and it could become more than that. It is integrated into the global economy and developing its own domestic sources of technological development, not just copycat industries but increasingly innovative and world-leading companies. China has already surpassed the United States in purchasing power parity, and it is projected to have the world's largest gross domestic product by as early as 2030. The last time the United States faced a competitor, or even a group of competitors, with greater economic power than its own was in the nineteenth century, before our own rise to global predominance. And when it comes to China's *potential* to generate even greater power, the United States has never faced a challenge of that scale in its entire history.

The challenge that China represents is amplified by the ways in which it is no ordinary nation-state. It embodies thousands of years of experience as the Middle Kingdom, a time when China viewed itself as the superior center of the world, surrounded by inferior states that it managed in a hierarchical tributary system. Indeed, the only exception to this nearly five thousand years of unbroken historical experience is what the Chinese call their "century of humiliation," the period from 1839 to 1949 when China was beset by civil war and dominated by foreign imperialist powers (the United States notably not among them). This century-long exception to a multimillennial rule just so happened to be the one blip in world history when the United States became the predominant global power.

It can be difficult, especially for Americans, to comprehend the sheer enormity of the shift in the global balance of power that is occurring with China's rise—or rather, its return. The only world that Americans know, the world of our own dominance, is one that few Chinese would recognize or accept as the natural order of things. This disjunction alone does not mean the two countries are "destined for war," but, as Graham Allison has written, it is bound to create tensions.[6]

It can also be difficult for Americans to appreciate fully the extent to which the leaders of China's ruling Communist Party are, in fact, ideologically motivated communists. This may have been debatable in the past, but it is harder to argue since Xi Jinping became president in 2012. Xi, who has set himself up as the ideological torchbearer of Mao Zedong, has led a renaissance of communist orthodoxy in government and culture, strengthened party control over businesses and the military, sidelined or purged many of his political rivals, and consolidated power in his hands more than any Chinese leader since Mao himself.[7] Indeed, after the removal of presidential term limits from China's constitution in 2018, Xi is poised to rule China indefinitely.

For Xi and other members of the ruling class within the Chinese Communist Party, the so-called princelings, communist ideology does not appear to be some relic of a bygone era but instead appears to be a meaningful motivation for their behavior and source of their official state policies. Leaked Communist Party documents suggest that the people ruling China, starting with Xi himself, are deeply paranoid and ideologically hostile to all forms of liberal influence. They see Western notions of constitutional democracy, human rights, free journalism, civil society, and open dissent as weapons that Western powers—above all the United States—use to weaken the Communist Party and China, which party leaders view as one and the same thing.[8] "Without the conspiracy of Western liberalism," Australian journalist-turned policymaker John Garnaut has written, the Chinese Communist Party "loses its reason for existence. There would be no need to maintain a vanguard party. Mr. Xi might as well let his party peacefully evolve."[9]

In this way, it is the ideology of the Chinese Communist Party that makes China's inherently unobjectionable desire for advanced technologies so troubling. Since becoming president, Xi has mobilized China in an unprecedented and comprehensive pursuit of 5G communications networks, biotechnology, artificial intelligence, and other advanced capabilities. Xi and other Chinese leaders seem convinced that these new technologies will enable China to "leapfrog" the United States and become the world's preeminent power. Indeed, under Xi's rule, Beijing sees advanced technologies as inextricably linked to China's national identity and the Communist Party's sense of destiny in restoring China to its rightful place at the center of world order.[10]

China's ambitions to become the world's technology superpower have been spelled out in a recent series of sweeping national strategies and industrial policies. *Made in China 2025,* issued in 2015, seeks to establish China as a world leader by 2025 in ten high-technology industries, including robotics, aerospace manufacturing, biotechnology, and advanced communications and information technologies, such as 5G networks. The *National Innovation-Driven Development Strategy* seeks to make China the world's "innovation leader" in science and technology by 2030, ranging from microelectronics and nuclear power to quantum technologies and space exploration, among other industries. In July 2017, China issued its *New Generation Artificial Intelligence Development Plan,* the goals of which Eric Schmidt, former CEO of Google and chairman of Alphabet, described this way: "By 2020, they will have caught up. By 2025, they will be better than us. And by 2030, they will dominate the industries of AI."[11]

What makes these national plans and industrial policies significant is their breathtaking scale and urgency. The US government, for example, has issued its own artificial intelligence plans. The Obama administration did so before China, and the Trump administration did so after. But these documents are more aspirational than directive. They were written largely by midlevel US officials. And they have done little to meaningfully shift US government investment. In contrast, China's plans are top priorities at the highest levels of the Chinese Communist Party and the People's Liberation Army, especially for Xi himself.[12] And they are backed by hundreds of billions of dollars of state-led investment.

What makes the Chinese Communist Party's technological ambitions even more threatening to the United States is a major way that Beijing enacts them—through a systematic global campaign to capture the world's best technology by whatever means necessary, which includes a massive foreign intelligence operation to steal trade secrets and intellectual property through cyber espionage and human spying. It also includes pressuring foreign corporations that seek to open operations in China—either as manufacturers or to sell to Chinese consumers—to hand over their intellectual property to the Chinese state as the price of doing business there. It includes the coercion of Chinese students at foreign universities, including at US universities, to spy on their peers, steal their research, and transfer it to the Chinese government. And it includes the deployment of significant sums of capital to invest in high-tech start-ups in Silicon Valley and elsewhere, establish ownership positions, and send their intellectual property back to China.

Even more worrisome are the unsettling ends to which the Chinese Communist Party is putting advanced technologies. What started with the "great firewall," an elaborate project to restrict the free flow of information into China, has developed into an all-encompassing and dystopian form of techno-authoritarianism by the Chinese state. A nationwide system of online monitoring and surveillance cameras, enhanced with artificial intelligence and facial recognition, oversees everything that Chinese citizens say, do, write, and buy, both online and in the real world. All of this personal information is used to build a comprehensive "social credit" system, in which the Chinese state rates the patriotism, loyalty, and adherence to government policy of every citizen, like an Uber rating system for human virtue as defined by the state. Those who "score" well would be rewarded with access to credit, government services, scholarships, better schools, and other perks. Those who do not score well could be denied these benefits. This is the most determined and expansive effort in human history to use technology to perfect dictatorship.

The military implications are also significant. All of the intellectual property that the Chinese Communist Party acquires or steals abroad, all of the breakthroughs that its domestic technology companies make, all of the joint ventures that US companies and research institutes conduct in China—all of it has a good chance of directly benefiting the People's Liberation Army under China's doctrine of "military-civil fusion." That is the law. And the goal, as Xi has stated, is to "complete national defense and military modernization by 2035 and fully build the people's army into a world-class military by the middle of the century." That military must focus on "preparing for war and combat," as Xi told senior military officers in 2018 while wearing fatigues and combat boots.[13] So, when US technology companies refuse to work with the Department of Defense but then do business in China, the practical effect is denying technology to their own military while providing it, knowingly or otherwise, to China's military.

The Chinese Communist Party's aims are evident not just in Xi's rhetoric and its legal doctrines governing the private sector but most alarmingly in its breathtaking military expansion. China's military budget has increased by 400 percent since 2006,[14] and though Beijing does not reveal the exact amount of its annual military spending, that number will surely continue to grow along with

China's GDP. This money is buying far more than what Chinese planners call "counter-intervention forces," such as dense layers of air defenses and advanced missiles of all ranges to deny US forces the ability to operate in the international waters and airspace of the Asia-Pacific region. The Chinese Communist Party is also investing more in advanced capabilities to project military power, such as expanded marine forces equipped with capable amphibious ships that could facilitate island seizures, as well as modernized fighter and bomber aircraft armed with ship-killing cruise missiles and other advanced weapons.

The centerpiece of the Chinese Communist Party's military buildup is the Chinese Navy. With an estimated 400 ships and submarines, the Chinese Navy is already larger than the US Navy, which currently consists of 288 combatants. Between 2015 and 2017, Chinese shipyards launched twice as many tons worth of naval vessels as their US counterparts.[15] China is turning out nearly a dozen new ships per year toward its goal of putting to sea a total fleet of 550 ships and submarines in ten years.[16] Many of these new ships, such as the Luyang-class guided missile destroyers and the Jiangkai-class guided missile frigates, are as capable as their US equivalents. And many of these new naval forces carry weapons, such as supersonic, sea-skimming cruise missiles, that are as good or better than those in US arsenals.

With this growing capacity to project military power, it is difficult to imagine that the Chinese Communist Party's ambitions are confined to China's borders, and that does not appear to be the case. China's government exports advanced weapons and the tools of high-tech authoritarianism to aspiring police states that want to surveil their citizens, regulate their thoughts, and crush dissent. It is using bribery, corruption, and other forms of coercion to interfere in the domestic affairs of other countries, including the United

States and its closest allies. It is accelerating its development of advanced weapons that are designed to push the US military out of critical areas of Asia, areas on which countless American jobs and much of our prosperity depend. And for two decades it has consistently expanded into the parts of Asia that China asserts belong to it and no one else. The Chinese Communist Party's appetite appears to be growing with the eating, and it is unclear where, or whether, that appetite will end.

Most Americans have spent their entire lives in a world defined by US military dominance. The United States has been able to project power anywhere in the world, penetrate into the physical space of any state adversary, and impose our will through military force or the threat of it. The idea that a time might come when America would not be capable of dominating an opponent in this way has been so unthinkable to us that we literally stopped thinking about it.

That time, however, may be coming. The rise of China is no ordinary foreign policy challenge. It is an unprecedented world historical event, and barring a collapse of Chinese power, China will emerge as a peer of the United States that achieves technological and military parity with us. That strategic reality will inherently lead to an erosion of US military dominance and a relative decline of American military power. It has rarely been the case historically, aside from those extreme instances of all-encompassing conflict, that a great power has been able to project force into the immediate space of a peer competitor and impose its will militarily. If a similar dynamic develops between the United States and China, as is likely, Washington would have to adapt to a world in which it no longer possesses military dominance over China.

What compounds this challenge is that Americans have been slow to recognize how the character of war has changed in recent decades. War is a perpetual contest between offense and defense.

The development of precision-guided weapons, beginning with the Assault Breaker initiative, gave the United States a decisive offensive advantage after the Cold War, because America's military alone possessed these capabilities. But as precision strike weapons have proliferated, especially in China, which Andrew Marshall and others foresaw as early as 1992, the advantage has shifted from the offense to the defense. It is extremely difficult to move a small number of large platforms halfway across the world and dominate a great power in its own backyard. The problem for the United States is that we have been building our military to project power and fight offensively for decades, while China has invested considerably in precision kill chains to counter the ability of the United States to project military power.

The greater danger now is where things are headed—which is toward the continued erosion of not only US military dominance but also America's ability to deter conventional war with China. If that deterrence disappears, what would likely fill the vacuum is a Chinese form of military dominance over much of the Asia-Pacific region—a region that is home to some of America's closest allies and that is the center of the global economy, on which the jobs, security, and well-being of millions of Americans depend. If that were to happen, Americans will be living in a world where the ultimate check on China's military or other ambitions is either the prudence and magnanimity of the Chinese Communist Party—or the willingness of the president of the United States to escalate every crisis with China, no matter how big or small, to the brink of nuclear war, which is hardly believable and even less desirable.

The stakes of this emerging strategic competition with the Chinese Communist Party are nothing less than what kind of future world we want to live in. This competition will require the full mobilization of our society, our economy, our diplomacy, our values, and our allies who share them. But the foundation for all of this is America's hard power, because the only way to ensure that this competition stays peaceful is by clearly being capable of defending what is most precious to us if the Chinese Communist Party—or anyone else, for that matter—chooses to confront us through aggression or violence. And that is what most concerns me: The entire basis by which the US military understands events, makes decisions, and takes actions—how it closes the kill chain—will not withstand the future of warfare. It is too linear and inflexible, too manual and slow, too brittle and unresponsive to dynamic threats, and too incapable of scaling to confront multiple dilemmas at once. That is why there is a growing concern within our defense establishment that America could lose a future war against a great power such as China.

This, to me, is something worse than change. Most Americans have lived blissfully free from the many kinds of privation, injustice, aggression, and depredation that countries through history have suffered at the hands of more powerful rivals that realized they could prevail in war if push came to shove. I have no desire to see how dangerous the future could become for Americans if we lose the ability to deter conventional war against the Chinese Communist Party or any other competitor. This situation should compel us to build different kinds of military forces that can defend Americans and our core interests in the absence of military dominance. This is possible, but it requires us to reimagine the kill chain and compete more urgently in the new strategic race over emerging technologies that is now under way.

SIX

A DIFFERENT KIND OF ARMS RACE

Since 2014, representatives of more than eighty governments have been meeting to consider an international ban on the development and use of lethal autonomous weapons, which the Department of Defense defines as machines "that, once activated, can select and engage targets without further intervention by a human operator."[1] These kinds of systems could be sent on military missions, identify targets using their own narrow artificial intelligence, and strike those targets, all without direct human control. They are weapons, in other words, that could close the kill chain without a human "in the loop." Opponents of these weapons, who are pushing to ban them, use a different name: killer robots.

In April 2018, the Chinese government expressed support for an international agreement "to ban the use of fully autonomous lethal weapons systems." The devil, however, was in the details. China explicitly did not call for a ban on *developing* these advanced weapons, and its official statement of policy defined lethal autonomous weapons incredibly narrowly—so narrowly, in fact, as not to capture the many ways in which the People's Liberation Army appears to be prioritizing the development of the very kinds of systems it says it never wants to use. On the same day that China called for banning the use of autonomous weapons, its air force announced a project to develop fully autonomous swarms of intelligent combat drones.[2]

This is in keeping with years of China's military development efforts and writings that envision a future of "intelligentized" warfare, most recently its "Defense White Paper of 2019." Chinese military planners have described their desire to create a "multidimensional, multi-domain unmanned combat weapons system of systems on the battlefield."[3] That translates to robotic combat systems everywhere—in the air and in outer space, on land, and at sea, including a potential "underwater Great Wall" of autonomous submarines. Although some details of China's military uses of advanced technology are known, such as experiments involving swarms of more than one hundred autonomous fixed-wing aircraft, it is unclear how far the Chinese military has progressed in its pursuit of intelligentized warfare. But if there was any doubt about the Chinese Communist Party's likely ambitions, China's Military Museum features a depiction of an aircraft carrier being overwhelmed by a "swarm assault" of unmanned combat aircraft.[4]

Some of these systems were on display in a major military parade that the Chinese Communist Party held in Beijing in October 2019 to mark its seventieth anniversary in power.[5] First came the DR-8, a supersonic drone believed to play a critical role in providing targeting data for China's "carrier killer" anti-ship ballistic missiles. Next came the Gonji-11, or Sharp Sword, a stealthy, unmanned combat air vehicle with an internal bay for weapons storage that also happens to be the spitting image of the US Navy's X-47B, which was prematurely retired in 2015. That was followed by DF-17 hypersonic missiles and a pair of unmanned submarines that could be part of that underwater Great Wall. It is doubtful whether any of these Chinese systems are lethal autonomous weapons, but it would be impossible to tell by looking at them. What would turn them into such weapons would just be some invisible lines of computer code.

The rush by many militaries to develop increasingly autonomous weapons has sparked concerns of a "global arms race for killer robots."[6] And that is just the tip of the iceberg. There has been talk for years about a growing "cyber arms race."[7] More recently, concerns have mounted over an "AI arms race,"[8] a "hypersonic arms race,"[9] a "5G arms race,"[10] a "quantum arms race,"[11] a "gene editing arms race,"[12] and a "new space race."[13] Whereas Russia features somewhat in these arms race apprehensions, China is the primary focus, because its economic and conventional military power are so much more significant and its intentions are so much less clear.

It is true that many Chinese actions are consistent with those of a rising power seeking to maximize its own influence. The problem is that many Chinese actions appear equally, if not more, consistent with a paranoid communist government that wants to perfect a totalizing form of dictatorship at home, export that model abroad, and make more of the world safe for its illiberal values of authoritarianism, crony capitalism, and surveillance states. And not only does the Chinese Communist Party see advanced technologies such as artificial intelligence as a big opportunity to realize its ambitions, it also sees the United States as the single biggest obstacle to those ambitions—not simply because of what America does but also because of what it represents as the world's most powerful embodiment of Western liberalism.

So, what does the Chinese Communist Party want? What is it building a "world-class military" to do? Do we think that it will not seize every opportunity advanced technologies afford to build the most capable weapons it possibly can? Do Americans trust the

Chinese Communist Party enough to live in a world where it has more capable weapons than the United States and our allies do?

Many Americans do not. And this is why the United States now finds itself in a new security competition over advanced technology, most consequentially of all with the Chinese Communist Party. Many Americans increasingly feel deeply uncertain and mistrustful about the intentions and ambitions of an illiberal competitor that could soon be as economically and militarily powerful as the United States—feelings, it should be noted, that are reciprocated in Beijing, despite long-standing and bipartisan US efforts to assuage China's security concerns.

Many defense and technology experts reject characterizations of this competition as an arms race.[14] They point out, correctly, that emerging technologies can transcend winner-take-all rivalries and produce mutually beneficial outcomes, even among great-power competitors. They also note, again correctly, that most applications of technologies such as artificial intelligence will go far beyond weapons. These capabilities are what defense and technology expert Michael Horowitz has called "enabling technologies,"[15] and they will enable sweeping social and economic changes that could transform how, where, and with what many people live and work. Even the defense applications of these technologies will not be limited to building arms. They will also enable new approaches to military logistics, health affairs, human resources, and many other non-warfighting functions.

All of this is true, of course. But practically speaking, there is a reason why people are alarmed about a new high-tech arms race, and the key word is *race*. Emerging technologies will certainly have broad non-military impacts, but they will transform the character of military power as well, and whichever actor is first able to develop and harness the military applications of these technologies could gain decisive strategic advantages over others. Americans must see this challenge clearly. After all, do we think there is any doubt in the minds of the rulers of the Chinese Communist Party—or Vladimir Putin, for that matter—that *they* are engaged in a new arms race with *us*?

This is not the result of technology. The origins of security competitions are always the result of geopolitics. These competitions are caused by the mistrust that exists between great-power competitors that are developing advanced technologies and their concerns over losing military advantage. This is why America and China are now locked in a strategic competition over emerging technologies that will be unlike such contests of the past in some key respects.

Previous arms races have been confined to the means of military action, from battleships to missiles to nuclear weapons. And that will be a feature of the current military competition, to be sure. Some emerging technologies will absolutely be used to build more and better conventional arms, such as advanced missiles and other weapons, and the United States and China will race to acquire them.

But there will be something unprecedented and different about this competition. It will also be a race to acquire new enabling technologies: artificial intelligence, quantum information systems, biotechnologies, and new space technologies. And though they will enable many human activities, both military and non-military, these technologies will also enable every phase of the kill chain—not only how militaries act but also how they understand and make decisions. The main goal will be accelerating the ability to close the kill chain and break rivals' ability to do so. And the main impact will be on what militaries refer to as command and control, the means by which human intent is turned into military effects on the battlefield. This competition, in other words, will have less to do with arms than with cognition. It will be a race over information. It will be a different kind of race, but it will still be a race for military advantage.

Turning this competition to our favor requires a better understanding of its underlying character, starting with the fact that certain aspects will, in fact, resemble a classical arms race. One area where this is already evident is hypersonic weapons, which are missiles or air vehicles that have entirely different flight characteristics from those of traditional munitions. Until now, weapons could fly in one of two ways. They could fly fast but relatively predictably, or they could fly relatively slowly but unpredictably. Ballistic missiles are fast but predictable: They largely travel along parabolic trajectories that are determined by gravity. No matter how fast they fly, their path of flight and point of impact can be predicted once they are launched. That is how militaries can shoot them down. Conversely, cruise missiles have traditionally flown slowly (for missiles, that is) but unpredictably: they can maneuver like airplanes, which makes it harder to know where they are headed, but because they fly slower, militaries can shoot them down, as well.

Hypersonic weapons are different because they can travel both fast *and* unpredictably. It is not clear where they are headed once launched, and they can fly so fast that there is very little time or ability to react. How fast? Upward of five times the speed of sound—that is more than 3,800 miles per hour, or roughly one mile in less than one second—and possibly faster. At these speeds, weapons could travel between China and Guam, the location of the largest US military base in the Pacific, in about thirty minutes. Indeed, US officials have stated that the United States currently has no capability to defend against, or even track, these weapons in flight.

Commercial technology companies likely will not be the source of hypersonic technologies for the US military. These capabilities will be built primarily by traditional defense companies, which have the expertise to develop these systems. But because Washington did not sufficiently prioritize hypersonic weapons, US officials now openly concede that China has gotten ahead. Beijing has poured billions of dollars into the specialized wind tunnels and other costly infrastructure that are prerequisites for developing hypersonic weapons, and it has reportedly conducted multiple test flights each year for several years in a row.

The United States is racing to catch up, but it will take a lot of time and money to develop and field hypersonic weapons. Just test firing a single missile can cost more than $100 million. This will limit the number of hypersonic arms that Washington, Beijing, and others will ultimately acquire. Indeed, the enormous impact of these weapons will be limited by the fact that militaries will likely be reluctant to expend their limited arsenals of them on anything but the most important strategic targets. And yet, the United States and China will race to develop and stockpile these weapons because they will be essential to maintaining conventional deterrence, and because neither will want to cede this important military advantage to the other.

The race to develop new high-speed weapons—not just hypersonic weapons, but also supersonic cruise missiles, electromagnetic railguns, hypervelocity projectiles, and new long-range cannons to fire them—is driving a related race to develop counters to these weapons. This competition will also have the hallmarks of a traditional arms race, and much of the focus is on directed energy weapons, such as lasers and high-powered microwaves. This technology has been an aspiration for a long time, and the US government has spent billions of dollars over several decades trying to develop it, often with little to show. Indeed, the joke about directed energy weapons is that they have been only five years away—for the past twenty-five years.

That, however, seems to be changing as well. New fiber lasers are much improved over older chemical lasers. Their beams are more concentrated and powerful. Low-kilowatt lasers can now burn holes through drones or vehicle engines. And higher-kilowatt lasers are being developed for defense against aircraft and missiles. The advantage of these weapons is their ability to shoot more often, much faster, at lower cost, and without the burden of resupplying ammunition. A traditional missile, for example, could cost upward of $1 million per weapon or more, and the even more expensive platforms that carry them eventually run out of weapons to shoot. Directed energy weapons, on the other hand, could cost as little as a few dollars per shot, and with a sufficient source of power, they could fire an indefinite number of times.

The big hurdle that remains is power capacity. It takes a lot of power to fire directed energy weapons of any strength. This means that they will be fielded first where power is most plentiful, such as land bases, large vehicles, and nuclear-powered ships, including aircraft carriers. It also means that directed energy weapons will initially be used less in offensive than defensive roles, such as protecting critical military installations from drone attacks. At present, these weapons are like early firearms, somewhat clumsy and of limited use now, and the race to build effective counters to these systems, such as mirrors and heat shields, has already begun. But for the first time ever, directed energy weapons are actually operationally viable, and eventually these systems could be fielded on military platforms of all kinds.

As militaries come to be more defined by digital features, such as data, software, computers, algorithms, and information networks, they will place greater value on weapons that can launch digital attacks. A shadow cyber war has raged for years, especially between America and its great-power rivals. It is a military competition that has many of the hallmarks of a traditional arms race. Much about this contest, such as cyber surveillance and theft, is not new. It is why all of my personal information from my security clearance investigation currently resides in China. Their hackers stole it from the US Office of Personnel Management in 2015, along with similar information on 19.7 million other Americans. The broader significance is that the cyber domain and the electromagnetic spectrum will be central battlegrounds of future war.

Take the F-35 Joint Strike Fighter, for example—the so-called flying supercomputer. It contains more than eight million lines of computer code that run its advanced digital systems. So, is the F-35 a piece of hardware that contains a lot of software? Or is it a digital warfare system wrapped in an aircraft? That this is even debatable shows the pervasiveness of the information revolution. It also points to the inherent cyber vulnerabilities of modern military systems and the attraction to cyber arms to exploit them. A major threat to the F-35 is not just enemy missiles but also the possibility that it could be cyberattacked before it ever gets off the ground.

The application of artificial intelligence will open a whole new front in the cyber arms race that focuses on the corruption or poisoning of data. Artificial intelligence is only as good as the data that trains its algorithms. This puts a premium on the integrity of that data, lest machines are trained to do the wrong things. For example, if a military is training algorithms to identify enemy tanks, and an adversary compromises that data and directs the algorithms to confuse tanks with buses, the result could be disastrous: an autonomous machine that is highly capable of doing the exact opposite of what humans intend. This is doubly worrisome because artificial intelligence does not need extra help to become less reliable: for as much progress as the technology has made and as fast it is improving, it still remains rather brittle, unpredictable, and unreliable.

Whether the weapons in question are digital, hypersonic, or directed energy, they will be arms—means of military action—and the United States and China will compete to amass these weapons in many of the same ways that great powers once competed over battleships and ballistic missiles. That competition could also encompass nuclear arms, especially considering how much China has been emphasizing nuclear forces in its military modernization. That challenge will have less to do with new technologies than with managing strategic competition between great powers.

What will make this contest a different kind of race, however, are the new enabling technologies that will revolutionize the entire kill chain, many of which are being pioneered in America by commercial technology companies. What these technologies will have in common is their ability to reduce the significant cognitive load on military commanders as they seek to understand events, make decisions, and take actions during high-stakes operations, where the sheer volumes of information and levels of complexity involved could quickly exceed human capacities to process them. The biggest impact of these enabling technologies could therefore be on how militaries command and control their forces on the battlefield—the ability of human beings to make more, better, and faster decisions that affect larger areas of physical and digital space.

One set of enabling technologies that could have this broader impact is quantum information systems. Quantum sensors, for example, could illuminate the battlefield better and generate unprecedented understanding for the militaries that possess them. As classical computers reach the physical limits of their power, quantum computers could become vital to processing all of the data that intelligentized militaries create and collect. And as traditional forms of encryption are threatened, quantum-resistant encryption could become indispensable.

In September 2019, news leaked that researchers at Google had demonstrated the first instance of "quantum supremacy," the moment when a quantum computer can perform operations that are beyond the practical limitations of classic computers.[16] The researchers claimed that their quantum computer could perform a certain operation in two hundred seconds that it would take a classic supercomputer ten thousand years to complete. It is unclear just how much of a breakthrough was actually achieved, and quantum researchers at IBM quickly took issue with Google's findings.[17] But those disputes, although important, should not distract from the larger point: the moment of quantum supremacy is drawing near. The bigger question is whether Google would allow its quantum computer to solve problems for the US military.

For all of their revolutionary potential, quantum information technologies that can be used to solve real-world problems are still a long way off. That means the race is now between those seeking to build quantum computers that can crack traditional encryption and those seeking to build new forms of quantum-resistant cryptography. It also means that when quantum systems do arrive, there likely will not be very many of them. A quantum computer is so complex, exquisite, and expensive that the United States and China may each have only a few of them. Such a system could be relevant for analyzing large volumes of historical data and training advanced algorithms, but its operational and tactical applications may be more limited.

An enabling capability that may have more immediate military impacts is biotechnology, which is already unlocking better understanding of human genetics and enabling the creation of customized treatments and technologies to augment human capabilities. This sounds creepy, but human performance assessment and enhancement are already common in the US military. Elite special operations units, for example, regularly use these kinds of technologies to assess candidates and identify the ones who are predisposed to bearing the cognitive and physiological loads of close-quarters combat, where individuals must understand, decide, and act precisely and repeatedly in split seconds. Similarly, enhancing human performance has become routine in the US military. Pilots, for example, regularly take modafinil and dextroamphetamines, central nervous system stimulants, to remain sharp and better able to understand, decide, and act during long flight operations.

Developments in biotechnology will mostly be extensions of current practices, albeit significant extensions. They will enable militaries to determine with far greater precision which individuals have won the genetic lottery when it comes to succeeding at particular military tasks, especially the command and control of large numbers of military forces under highly dynamic conditions. Of specific interest will be identifying those elite few who are more cognitively and physiologically capable of understanding, deciding, and acting effectively as the volume and velocity of warfare increase. Biotechnology will make possible customized medications that enable individuals to be significantly better at what they are predisposed to doing well. Performance enhancement might even include tailored biological augmentations that enable military operators to command and control military forces operating at machine speeds.

Enhancing human performance might also extend to the augmentation of human beings with intelligent machines. Brain-computer interface technologies, for example, could enable humans to supplement their own cognitive abilities with computers that perform more menial military tasks, such as processing overwhelming amounts of sensory information and generating better understanding of highly dynamic events. Similar technologies might serve as highly capable digital assistants, enabling human commanders to recall necessary information rapidly and right when they need it, receive updates on important events that are beyond their present focus, and make better decisions based on the recommendations of intelligent machines that have a fuller, more accurate picture of events than they do. This kind of biotechnology allows machines to do what they do better than humans, thereby enabling humans to focus on what they must do, which is command and control the conduct of warfare and the closure of kill chains.

In the competition over biotechnology, it is hard to believe that the United States would cross certain ethical lines, but it is less clear whether the same can be said of the Chinese Communist Party. It has already turned the Xinjiang Uighur Autonomous Region into what the United Nations has called "something resembling a massive internment camp" for the minority population there.[18] It has allowed researchers to produce the world's first genetically edited human babies.[19] It condones genetic experimentation on animals, including non-human primates, that is far more restricted in the United States. It is not a significant moral leap to imagine China genetically engineering superhumans who are optimized for certain military tasks or developing precision biological warfare agents that, like designer drugs in reverse, infect specific groups of people or individual members of a rival military. Indeed, biotechnology could be one area, more than others, where values differences between America and China have the greatest military ramifications.

Another set of enabling technologies that will transform every aspect of the kill chain are new space capabilities. A competition akin to a new space race is under way between the United States and

China, evident in the rush to blanket the earth with thousands of small satellites that provide everything from high-speed communications to high-resolution intelligence. New space-based capabilities will be central to how militaries command and control their forces. Even more than today, future kill chains will flow through space, enabling militaries to distribute the process of understanding, deciding, and acting across large networks of systems rather than depending on single platforms to close the kill chain on their own. The result will be a dramatic expansion of "time-sensitive targeting": the ability to find moving targets, track them, and strike them before they have moved away. This type of targeting is rare today, largely because of a scarcity of satellites.

The proliferation of satellites, however, is only the beginning of the new space race. Spacecraft have always been limited by the impracticality of refueling them. They have only as much fuel as they could carry into space, and when it is gone, they cannot actively propel themselves any farther. This has restricted spacecraft to orbiting Earth, but emerging space technologies are changing that. Indeed, the new space race will also be a competition to build the infrastructure off of Earth that enables and secures a spacefaring future.

These technologies are being developed now. In the coming years, it will be possible to service, assemble, and manufacture complex orbital infrastructure in space that would be impractical to launch from Earth. This could include vast space-based solar power arrays to capture more of the sun's energy than is possible on Earth, where our atmosphere absorbs or deflects it. Power-beaming technologies will transfer that energy around space. Space-based mining technologies could extract ice from the moon and asteroids, utilizing the underlying oxygen to fuel rockets and support human life in space. The means of production to support space operations will increasingly shift off Earth and into orbital bases and perhaps onto the moon, where spacecraft and other critical space infrastructure could be produced using 3-D printing and advanced manufacturing. It sounds like fantasy, but it is not.

In time, space will be transformed into a unique domain of human activity, and this will inevitably have military implications. Space operations in the coming decades will come to resemble maritime operations in the nineteenth century, when industrial age great powers built global networks of coaling stations and other infrastructure to project naval power in defense of their expanding commercial interests. A similar dynamic will occur in outer space, and the two states that will most shape humankind's spacefaring future will be China and the United States. It is hard to imagine their strategic and military competition will remain confined to Earth.

Of all the new enabling technologies, perhaps the most consequential from a military standpoint is artificial intelligence and machine learning. The most immediate impact of these technologies will be their ability to improve human understanding in warfare. They can enable human commanders to make better decisions based on available information. The US military is drowning in data. It uses powerful and exquisite sensors, all machines, to suck up oceans of information about the world, but then it leaves the job of making sense of it to humans. There simply are not enough people in the US military to interpret all of this data, nor will there ever be. As a result, most of the information that the US military collects either goes unused or is thrown away—a complete waste that results in people making less-informed decisions, often about matters of life and death.

Although artificial intelligence still cannot do many things well, interpreting sensor data is not one of them. Machine learning algorithms can be highly effective in consuming large volumes of data and then identifying patterns, objects, and trends within that information. For example, a computer vision algorithm that has been trained to identify people in images can rapidly review millions of photos and find the ones with people in them. Machines can perform this particular task better and exponentially faster than humans can. They can pick out images of humans that the naked eye simply cannot see. And their abilities are not limited to identifying people. Well-trained machines can find all different kinds of objects, sounds, or other signals within vast quantities of different kinds of sensor-gathered information.

Humans will eventually be able to delegate much of the cognitive burden of closing the kill chain to well-trained intelligent machines, thereby enabling people to focus on making better and faster decisions in warfare. In time, human decision making could be improved by machines that can predict and make informed recommendations about the most effective courses of military action. This is not the case with US war plans today, which are largely linear, static, and inflexible. As soon as those plans fail to survive contact with reality, commanders are left to improvise their next moves with limited ability to make sense of dynamic environments on relevant timelines and recompose their forces in new ways to take different actions other than the ones they were preplanned to take.

It is not hard to imagine that militaries will train artificial intelligence programs using hundreds of years' worth of simulated experience, much as AlphaStar learned to play StarCraft II, to determine the optimal ways to conduct military campaigns. Every possible event that could occur during an actual war would be an event the machine has confronted many times over in simulations. The machine could make informed recommendations to human commanders of decisions and actions they should take in any given contingency, as well as calculate the probability of their success. Human commanders would certainly be free to dismiss this computer-generated advice, but having it could give them access to better information and more considered options than they have now.

An even more significant development will be the emergence of intelligent machines—namely, systems that can understand, decide, and act on information independent of direct human control but still within the parameters that humans define. An intelligent machine would be capable, on its own, of making sense of its surroundings, navigating the environment, identifying relevant pieces of information that humans have trained its algorithms to find, and taking actions on the basis of that information. The complexity of those actions would be constrained by the restrictions humans impose and the technical limitations of the machine itself, of both its robotic hardware and its intelligent software. While it is certainly true that any lethal autonomous weapons would be intelligent machines, militaries will be able to use non-lethal intelligent machines for a multitude of purposes that do not involve violence, from intelligence gathering to automated logistics.

In this way, intelligent machines may be more important than 5G communications in the narrow sense of how states build military systems and use them to fight differently. Though 5G networks will be critical for broader economic and geopolitical purposes, communications networks are really just pipes for information. Wider pipes allow more information to flow through them faster, and 5G pipes will be the widest yet. But faster information flows might not necessarily change the way militaries operate. Intelligent machines will. They will interpret most of the information they collect independently, using artificial intelligence to identify critical pieces of intelligence within oceans of other data. Sending those small bits of pertinent information to humans or other machines will not require 5G communications networks. In fact, intelligent machines will likely transmit *less* information across the network than current military systems do—it will only be the important bits, rather than all of it. This will make it feasible for humans to remain in contact with intelligent machines, including during military operations.

When it comes to the military competition for artificial intelligence, China has one big advantage that Americans should not discount, and that is its enormous size. One of the most important development tasks of artificial intelligence is using recent breakthroughs, especially in deep learning, to deploy artificial intelligence at scale.[20] That will require enormous quantities of data and enormous amounts of computer processing. This kind of huge centralized operation is something that China's authoritarian system will likely do well. The Chinese state is accumulating data on its 1.3 billion people with little regard for civil liberties, as well as growing numbers of foreign citizens living in countries whose governments are fielding Chinese surveillance technologies, and it is racing to develop better computing. Indeed, since 2014, the Chinese state has deployed around $65 billion to build a domestic supply of semiconductors.[21]

Most Americans simply do not appreciate the scale of the advanced technology challenge that China poses, how rapidly it is catching up, and how it has already surpassed our own capabilities in some important areas. Nine of the world's top twenty internet companies are Chinese. The country is producing five times as many engineers as we do and is beginning to equal the United States in the skill of its researchers. More than half of the most-cited research papers on artificial intelligence in the world in 2018 were produced in China. And in recent years, Chinese teams have been winning major international competitions in facial and voice recognition.

China's technological progress extends well beyond advances in artificial intelligence. Its companies represent 80 percent of the commercial drone market. In 2017, Chinese scientists used a quantum communications satellite to make a Skype call from Beijing to Austria that was secured from end to end using quantum encryption. Two years later, China landed the first spacecraft on the far side of the moon, and it has announced plans to build a lunar base near the south pole of the moon, which is estimated to contain considerable deposits of ice, the oil of outer space. While disagreements between Washington and Silicon Valley persist, the Chinese Communist Party is directing the "military-civil fusion" of the People's Liberation Army and China's technology sector.[22]

In any arms race, even the different kind of race that confronts us, there are considerable inherent risks. States feel enormous pressures to develop and deploy technologies for military purposes, even in advance of clear policy and doctrine to govern their use, and perhaps in advance of the full scope of testing and training. This is a real concern with all emerging technologies, but especially with artificial intelligence, which still has a long road of further development. If this process is not deliberate, thorough, and precise, the result could be the deployment of intelligent machines that are unpredictable and fundamentally dangerous to their users, to say nothing of innocent civilians.

What is more unfortunate is that we probably cannot avoid or avert this new and different arms race. The main reason is an uncomfortable truth that everyone, especially advocates of arms control, such as those meeting in Geneva to ban the development and use of "killer robots," must confront: the vast majority of instances in history when nations have agreed to restrict their use of the most terrible weapons—from nuclear weapons and incendiary bombs to poison gas and biological weapons, land mines and cluster bombs—agreement has occurred only *after* those weapons were developed, *after* they were widely used in combat, or *after* nations determined the weapons were not as effective or beneficial as assumed. A main reason states agreed to ban mustard gas, for example, was because it had the unfortunate habit of getting blown back into the faces of the very soldiers who had employed it.

States may still pursue arms control agreements as a way of establishing international norms against certain kinds of weapons, but agreements involving artificial intelligence, robotic vehicles, and other emerging technologies might provide little to no practical assurance that militaries will not weaponize them. If a country could produce self-driving commercial vehicles, for example, it would not struggle to produce self-driving combat vehicles. If a country could produce a fully autonomous aircraft to perform non-lethal military missions, it could quickly and easily arm those aircraft and configure them to conduct acts of violence. Opposing forces facing these autonomous systems on the battlefield would struggle to differentiate lethal from non-lethal technologies. Even if states could verify that their rivals were not cheating on their commitments, that would tell them little about their opponents' actual capabilities.

The United States must also face up to the uncomfortable reality that we are unlikely to "win" this different kind of race, if winning such a contest is ever possible. If China continues to grow wealthier, more powerful, and more technologically advanced, it will likely come to possess many, if not all, of the military capabilities as the United States. In some areas, China may even surpass the United States. Realistically, the best-case scenario is not victory in this race but parity. That may not sit well with many Americans, who are used to dominating our military competitors, but the idea is nothing to scoff at. It is how war is prevented. The main purpose of building highly capable military forces and demonstrating their abilities to our rivals is to convince opponents that they have nothing to gain and everything to lose by using their own military forces against us.

The greater danger for the United States is failing to recognize the true gravity of the kind of military technology race with the Chinese Communist Party that we are facing and falling behind because of our lack of urgency to run it. From President Xi Jinping down, China's senior leaders are fully mobilized and moving with awe-inspiring speed to become the world leader in emerging technologies. They, too, seem to value the enabling potential of these technologies, and for Chinese leaders, the most important thing these technologies will enable is China's ability to "leapfrog" the United States and establish itself as the world's preeminent power.

The United States may not win this different kind of race, but right now we are at risk of losing it. We are at greater risk of moving too slowly and cautiously in this competition than moving too fast and recklessly. And perhaps the most consequential new military capability in that regard is intelligent machines, including the weapons that are now a focus of international debate in Geneva. The question is not whether the United States *can* build intelligent machines that are capable of closing the kill chain without humans in the loop. The bigger question is whether we *should*.

SEVEN

HUMAN COMMAND, MACHINE CONTROL

A barrel bomb is made from a fifty-five-gallon oil drum that is filled with gasoline, explosives, nails, glass, ball bearings, nuts and bolts, and other metal shrapnel. It has no guidance system. It is usually dropped or rolled crudely from helicopters, where it falls hundreds of feet by the force of gravity alone onto its targets. Barrel bombs exist for one reason only: to cause as much death and indiscriminate human suffering as possible at the lowest possible cost.

I first learned about barrel bombs in January 2013 on a trip with John McCain to the Zaatari refugee camp, a massive city of canvas tents sprawled across the sandy moonscape of northern Jordan that was then home to tens of thousands of Syrians who had fled the civil war raging in their country. It was there that we met a Syrian mother who had recently lost all five of her children in a barrel bomb attack by forces loyal to President Bashar al-Assad. She fled the country with nothing more than her own life. As a parent myself who had just had my second child, I was haunted by the emptiness in her eyes. So was McCain, who had seen more of death and suffering in war than anyone I have ever known. In the nearly one hundred foreign trips that I took with him, I cannot recall a single meeting that affected McCain more deeply.

"Fathers, brothers, sisters—they all move on after the loss of a child," McCain said to me afterward as we drove back to Amman. "But not the mothers. The mothers never move on."

I often think about that Syrian mother as I consider the serious ethical questions surrounding the use of intelligent machines in war. I imagine myself as an innocent civilian in a future conflict. I imagine fleeing my home with my children as military aircraft drop weapons all around us. I try to imagine, as best I can, being that afraid for my life and for the lives of my children. I imagine the feeling of not knowing whether human pilots or intelligent machines are operating those aircraft—and whether the means of delivering violence are crude and dumb (such as barrel bombs) or smart and accurate (such as precision munitions).

The bigger question I ask myself is whether, in that moment, I would care. Would I care why or how the military forces above me are deciding to use violence? Would I care whether human beings are making every decision and dropping those bombs or whether intelligent machines are doing it without direct human control? Would I care more about the character of the actor or the conduct of the action? Put simply, would I have any concern beyond the survival of my children and me by whatever means possible?

I also try to imagine how I would feel if I knew that the operators of those military aircraft—whether humans or machines—were more likely, for whatever reason, to kill my children and me than other operators that could have been in combat that day. How would I feel if the people who deployed those military operators to make life-and-death decisions did so knowing that they were more likely to kill civilians? Would I consider the most ethical decision to be anything other than that which is least likely to result in the loss of innocent lives, including my children's and mine?

The point here is not to argue whether to keep the kill chain firmly in human hands or to turn it over, in part or in whole, to machines. Rather, the point is that the current debate over the role of intelligent machines in war—one of the most serious ethical questions raised by emerging technologies—too often focuses on the wrong things. We seem overly concerned with means rather than ends, actors rather than actions, "killer robots" rather than effective kill chains. Ironically, even our ethical debates about military technology seem overly focused on platforms.

Many defense and technology experts, such as Paul Scharre, Heather Roff, and Joe Chapa, have written extensively on the serious ethical and technical challenges involved in the military use of intelligent machines—challenges that the deepening strategic competition between the United States and China is only exacerbating.[1] Artificial intelligence, at present, can be brittle, opaque, unreliable, unpredictable, and prone to error. It struggles to contextualize information. But as the technology becomes more capable, which is happening quickly, the possibility of relying on increasingly intelligent machines to perform military tasks will grow, as will the pressure to do so amid an intensifying race to develop faster, more effective kill chains.

I have grown frustrated with our debates on the military implications of intelligent machines. Many people are posing good questions, but it still feels as if there are not enough good answers to guide those in positions of authority who must make difficult decisions now about how to spend money, develop military systems, and govern how Americans in uniform will actually use and relate to this technology. I speak with a lot of policy makers, military officers, and researchers who privately suggest that the United States will, and should, develop highly intelligent machines for military purposes, including lethal autonomous weapons. But those same people rarely take such clear positions in public. I do not think this secrecy ultimately helps to develop good, practical answers to these ethical questions that Americans can support.

Intelligent machines will have a significant impact on the future of war, but for years and possibly decades to come, they will represent less of a qualitative break with recent technology than a shift, albeit a big shift, along a continuum that we have been moving along for some time. What we are really talking about is the ability of machines to perform specific tasks that so far only humans have been capable of. The key questions are not who or what is performing those actions, but whether human agency is clearly initiating those actions, especially the use of violence in war, and whether there is clear human accountability for the consequences of those actions. In this way, I believe that our ethical principles, laws, policies, and practices can incorporate most military uses of intelligent machines—if we think about them the right way.

The US military often refers to the relationship between human beings and intelligent machines as "human-machine teaming." I dislike this term because *teaming* suggests a relationship of equals. The better way to think about this relationship is with the military concept of command and control, which refers to the hierarchical relationships between people in military organizations. Superior officers command human subordinates to control the execution of military tasks. As machines become more capable of performing tasks that people now perform, our concepts of command and control can and should evolve. Humans must remain in charge and issue orders, but increasingly intelligent machines will be able to carry out more of those orders to enable human understanding, decisions, and action. Our challenge, then, will be adapting to an era of human command and machine control.[2]

When I worked for McCain, some of our most important allies were human rights groups that brought to light the atrocities that were occurring in Syria and other conflicts around the world. They helped to tell the stories of the victims and document the many evils that humans commit against their fellow humans in war. It always puzzled me, then, to see many of these same groups advocate so strongly to "ban killer robots." They would argue that machines should not be allowed to use violence because they are too cold, too unfeeling, too lacking in empathy, mercy, and other ethical qualities that humans possess. But the same groups would then publish case after tragic case in which humans themselves disregarded their own ethical qualities and committed unspeakable atrocities out of vengeance, fear, greed, and other passions that machines also lack. It is tempting to want to have it both ways, but alas we cannot.

Much of the understandable fear about intelligent machines involved in war arises because these systems conjure dystopian images of the Terminator or Skynet. But those kinds of self-aware machines are not what is currently at issue. We are not talking about machines with *superintelligence* or artificial *general* intelligence. We are not talking about machines that would be capable of overriding their own programming and making their own decisions without human limitations, influence, or control. Researchers think that level of machine intelligence could be decades away if it is even possible at all.[3]

The real issue now and for the foreseeable future is the military use of *narrow* artificial intelligence. Machines with this more limited artificial intelligence can perform specific but circumscribed tasks, such as identifying objects in images and navigating their surroundings. These abilities are remarkable achievements, and the tasks that intelligent machines will be able to perform will increase in number and complexity. Even then, this level of machine intelligence is a far cry from self-conscious machines that could freely violate rules that humans have established.

It is important here to make another key distinction: What machines are capable of doing (automation) and what humans permit machines to do (autonomy) are very different things. There is no such thing as an "autonomous" machine, technically speaking, because *autonomy* describes a relationship, not a thing. It is the relationship between a superior actor that delegates tasks to subordinate actors under certain parameters. This kind of relationship is pervasive in militaries. Indeed, it is the foundation of effective and ethical military conduct in war and peace.

Human commanders routinely grant human subordinates the autonomy to perform military tasks, including the use of violence, but they grant autonomy within certain parameters, only if they believe their subordinates are capable of making good decisions on their own within the constraints of the orders given. This comes down to three factors: training, testing, and trust. Human commanders rigorously train their subordinates to perform the tasks they plan to delegate to them. They test those subordinates, over and over again, to determine whether their subordinates can reliably, predictably, and effectively perform the tasks they are given. And it is through that repeated experience of training and testing that human commanders determine whether they can trust their subordinates with autonomy to act on their orders. In fact, commanders would be held accountable for failing to properly train and test their subordinates.

Accountability is a core component of trust. When military officers at war receive new troops under their command, the reason they should be able to trust those human subordinates to do their jobs effectively and ethically is because the commanders know that another human being has trained those troops and is accountable for deciding they were safe and ready to send to war. The same process of trust and accountability also applies to machines. When servicemembers are issued weapons, they are able to trust the safety and effectiveness of those weapons because other humans have tested them extensively to determine that they will perform as intended under the many conditions in which they might be used. If machines—or human beings, for that matter—perform in unsafe or ineffective ways in combat, the accountability for that failure lies with the trainer and tester, not the user.

Indeed, this process of training, testing, and trust building is how the US military has incorporated increasingly capable machines into its formations for a very long time. Antitank guided missiles, cruise missiles, and other "fire and forget" weapons, for example, can find their own way to the targets that humans have identified on the battlefield but cannot actually see. Advanced weapons such as these took over many of the tasks associated with understanding, deciding, and acting that human beings once had to perform themselves. And the US military decided to use these machines for these purposes only after long periods of training, testing, and trust building to determine and validate their capabilities and limitations.

This same framework can apply to machines as they become more intelligent and more capable of doing more of the jobs that previously only humans could do. Humans will have to develop trust in those machines before deciding to grant them greater autonomy, even to do tasks they may technically be capable of. Commanders will build this trust in machines in much the same way they have come to trust their human subordinates: through rigorous training and repeated testing. The machines must demonstrate to the humans who are accountable for them that they can safely, reliably, and effectively perform the tasks that might be assigned to them. And if intelligent machines cannot meet those high standards, militaries should not use them. Period.

This process of building trust and establishing accountability is especially important in a military context, where the consequence of giving jobs to untrained, untested, and untrustworthy subordinates can endanger human lives. That is why commanders never give their subordinates complete autonomy, even after they gain trust in them. Instead, they issue clear orders and rules that constrain their subordinates' freedom to act. The same paradigm would apply to intelligent machines that have demonstrated their ability to perform specific tasks through training and testing. Human beings would limit the degree of autonomy granted to machines by purposefully constraining them through orders and rules in their programming. Commanders will always want and need to limit the autonomy of their subordinates, regardless of whether they are human beings or intelligent machines.

Even into the future, certain tasks are so consequential that commanders will keep greater or even full control of the process of decision making and not entrust those decisions to machines, no matter how capable they become. The clearest example is nuclear command and control, where the risks of miscalculation or unpredictability do not get any higher. These are decisions where deliberation is especially vital, where speed is not advantageous, and where humans should remain in control of each stage of understanding, deciding, and acting. Nuclear weapons are the most significant example of a military issue over which human commanders should retain tight control over the entire kill chain and not rely excessively (or at all) on intelligent machines, but there will certainly be others.

For the many military tasks that do not fall into this category, however, the prospect of human commanders delegating more to intelligent machines opens a huge ethical opportunity: the opportunity to better differentiate between jobs that machines *can* do and jobs that humans *should* do. Human decision making has an inherent ethical value that derives from the human capacity to make highly complex decisions about right and wrong in different contexts. Machines cannot do this now and may never be able to do it well. The ethical value of human decision making is wasted when people do jobs that require them to use little of their ethical faculties. These are the kinds of technical tasks that too many members of the US military still have to do each day—tasks that involve driving machines from one place to another, transmitting information from one system to another, steering sensors to look and listen at things in the world, or sifting through oceans of data to identify and classify the relevant pieces of information.

As intelligent machines become capable of performing these kinds of technical tasks more effectively than humans can, allowing them to do so can liberate more members of the military to do work of greater ethical value. They can spend more of their days solving complex problems with other people, making operational and strategic decisions, contextualizing critical information, distinguishing between right and wrong, and commanding people and machines to perform critical missions. These are the kinds of jobs that Americans actually join the military to do. In this way, intelligent machines could enable more human beings to concentrate on the ethics of warfare than ever before.

For people to build greater trust in intelligent machines and delegate to them military tasks that humans are now performing, the machines should not have to be perfect—they should just be more effective than the people performing the tasks now. And in the US military today, those people are often scared and emotional twenty-somethings who work under incredibly stressful and unpleasant conditions. They are often distracted, tired, and confused. They have access to only a fraction of the information at their disposal because there simply are not enough people to make sense of it all. They are regularly making decisions based on limited and highly uncertain understandings of what is actually happening. These men and women are far from perfect. They make mistakes all the time. They are, in other words, human.

Current intelligent machines could perform many of these tasks more effectively than the people who are doing these jobs now. Machine learning cannot do everything well, but much of what it has been well trained and tested to do, such as rapidly identifying useful information in vast sets of data, it has repeatedly shown that it can do more accurately, more consistently, and far more quickly than humans. These machines do not get tired or miss things because they need to sleep or go to the bathroom. They can identify the information humans have directed them to find, thereby improving people's ability to understand events, make decisions, and act. Indeed, a principal function of machine learning is to help better inform and educate human beings, not to replace them in every instance.

It is not difficult to imagine that greater use of these intelligent machines today could lead to more ethical outcomes in war right now, such as reducing the number of civilian casualties or the risk to US troops. Indeed, the alternative could be greater civilian casualties and increased numbers of US troops killed in combat. We tend to think mostly about the risks of trusting intelligent machines too much. But not trusting them enough could also lead to some eminently unethical outcomes.

There are no ideal solutions when it comes to warfare. The use of violence inherently involves deep levels of uncertainty, risk, and unappealing trade-offs—something that both the opponents of and proponents for military use of intelligent machines must always remember. It is guaranteed that both humans and machines will err in the conduct of war. More intelligent machines will still make mistakes; they will just make different mistakes. Some of those mistakes could be less costly than those that humans make, but some could be more costly. This becomes especially problematic if, as opponents of autonomous weapons fear, humans use those weapons over time for more expansive purposes and in less proportionate and discriminate ways. This is a legitimate cause for concern, and it highlights the critical question at the center of our current debates over artificial intelligence and warfare—whether humans could ever train and test intelligent machines enough to trust them to close the kill chain without a person in the loop or directly in control.

I believe we can. In fact, we already have. But even framing the issue in this way does not readily convey the vast extent to which human choices and values are engineered into the deepest recesses of even those machines capable of acting with the greatest degrees of autonomy.

Lethal autonomous weapons have existed for a long time. Such systems, with varying degrees of capability, are currently in use by at least thirty different states. The US Navy, for example, has used the Phalanx gun and Aegis missile defense systems to defend its ships for decades. Though far less capable than the intelligent machines of today and tomorrow, these systems can be switched into a fully automatic mode that enables them to close the kill chain against incoming missiles without human involvement. The decision to trust those machines to do so was born of necessity: it was unlikely humans could respond fast enough to counter incoming missiles. That inability was deemed a greater danger than the option of turning the kill chain over to a machine that could shoot down missiles in time-sensitive situations more effectively than humans could.

The intelligent machines being developed today are far more sophisticated than the Aegis system, but even if they were permitted to close the kill chain on their own, they would not be choosing to do whatever they want. To the contrary, a machine's ability to operate autonomously is confined by boundaries that humans have established. Its ability to identify targets is limited and based on algorithms that humans have written, trained, tested, and come to trust as effective. The machine's ability to use violence against targets is further constrained by parameters that humans have defined, such as how much and how often the machine can fire. And the only way the machine is ever sent into combat in the first place is because humans have made the decision to release it from their control—a decision for which they are ethically and legally accountable.

That process of establishing accountability for the military use of intelligent machines would not radically differ from how human commanders choose to delegate control over the use of violence today, regardless of whether they delegate that task to subordinates who are people or machines, highly intelligent machines or more basic ones. The humans who would train and test an intelligent machine would be accountable for determining its capabilities and limitations. The more intelligent the machine and the more complex the task, the more training and testing would be necessary before humans can trust that machine to work as intended. Similarly, human commanders who must employ that machine would be accountable for using it consistent with its established capabilities and limitations.

There is nothing inherently unethical or illegal about lethal autonomous weapons that requires them to be banned outright. Indeed, the legality of weapons is a question that military ethicists have labored to answer for decades, and the conclusion forms a relatively clear basis in US and international law for determining what makes a weapon unlawful. First, the weapon cannot be indiscriminate by nature. This does not refer to how human combatants might use or misuse the weapon. It means only that the weapon itself cannot be specifically designed to cause indiscriminate harm. Second, a lawful weapon cannot "cause unnecessary suffering or superfluous injury." This rule is intended to exclude, for example, bombs filled with glass shards that an X-ray machine could not detect in the human body. Finally, a lawful weapon cannot cause harmful effects that are incapable of being controlled. The oft-cited example here is biological weapons, which spread harm beyond human control once released.[4]

These are the rules by which humans have judged weapons to be illegal for decades. The key determinants are the effects of the weapon, not the level of intelligence of the actor deploying it. It is possible that intelligent machines could be built in such a way as to violate one or more of these criteria. But as legal scholars Kenneth Anderson, Daniel Reisner, and Matthew Waxman have argued, "None of these rules renders a weapon system illegal per se solely on account of it being autonomous."[5] Indeed, an autonomous weapon is not inherently indiscriminate, predisposed to cause superfluous harm, or uncontrollable. To the contrary, what an autonomous system does, even if acting independently of direct human control, is limited and defined by what humans have programmed it to do. In this sense, there is nothing fundamentally new or unique about intelligent machines, even ones that can close the kill chain on their own, that would legally require them to be banned.

The more important consideration, of course, is what human combatants use weapons to do, and this almost always is situationally dependent. A machine that can find and attack targets without direct human control seems scary in the abstract. If such a machine were turned loose to hunt and kill innocent civilians, it could rightly be called a killer robot. But if that machine was neither inherently unlawful nor disproportionate to the military task at hand, and if it was used to protect human lives—indeed, if it could keep US troops safer than any other weapons they have—would it not be ethical to use that weapon for that defensive purpose? Indeed, would it be ethical to send US troops into harm's way *without* this means of self-defense?

Just because lethal autonomous weapon systems are not inherently illegal does not relieve human commanders of the responsibility to develop the same level of trust in their safety and effectiveness as any other machines—or fellow humans, for that matter—that they choose to send into combat. After all, lethal autonomous weapon systems have made costly mistakes before. On more than one occasion, for example, US Patriot missile systems operating with a high degree of automation have shot down friendly military aircraft. Incidents of fratricide are sadly not uncommon in war. Machine error has caused some of these tragedies. Human error has caused many more. There is no getting around the fact that mistakes, including catastrophic mistakes, are an unfortunate reality of warfare.

Neither humans nor machines will ever be perfect. There is always the risk that people and machines, when granted greater autonomy, will not perform as their human commanders intend. Commanders must determine how best to mitigate this risk and increase their trust in subordinates to do their jobs safely, predictably, and with the fewest mistakes possible. Here, too, that trust will be built, as it always has been, through training and testing, and these processes apply equally to human beings and intelligent machines.

A complicating factor is that the decision-making process of current intelligent machines can be highly opaque. The classic case is of the theoretical house-cleaning robot that concludes the best way to keep the house clean is to lock the family in the basement—the wrong decision, but not necessarily illogical from the machine's perspective. Similarly, artificial intelligence programs that have mastered games such as chess and Go eventually achieve a superior level of gameplay where the reasoning behind their moves and strategies is seemingly inexplicable to their human creators. For this reason, researchers are already working to develop more explainable artificial intelligence, machines that could reveal the reasoning behind their decisions and actions. Being able to follow the reasoning process of intelligent machines will not only help improve human trust in them but also make those machines more effective.

But how much of this really matters when we decide whether a machine is safe, effective, and ethical to employ? Militaries have been sending people to war forever despite the fact that human decision-making processes can be opaque, their ability to recall events through the fog of war can be suspect, and they may lie about the actions they took and why they took them. Whether commanders can know what people or machines are thinking and why they make the choices they do is not the basis for deciding to send them to war. Commanders send people or machines to war because they have consistently shown through training and testing that they can successfully perform the tasks delegated to them. And that is the standard by which they are judged in war—whether they acted effectively and ethically under the circumstances. When it comes to the use of violence in combat, whether by people or machines, the ethics of warfare is more concerned with the specific actions taken rather than the motivations of the actors.

It is for the very purpose of limiting specific, potentially unethical actions in war, such as the killing of civilians, that the United States and other nations have developed laws and rules to constrain human actions in combat. US and international laws seek to limit the use of violence in war with the law of targeting, which has three criteria. The first is discrimination, which requires combatants to take all reasonable steps not to attack civilians or civilian objects. The second is proportionality, which requires combatants to evaluate the military gains of an operation in light of the civilian losses that could occur as a result. The third criterion requires commanders to take "precautions in attack" that spare civilians from unnecessary suffering.

Human commanders rarely leave it entirely to their subordinates' discretion to determine how to follow the laws of targeting. Instead, commanders establish clear rules of engagement to limit how and under what circumstances their subordinates can use violence in combat. They may forbid their troops from firing unless fired upon, for example, or permit them to fire only if the expected civilian losses are below a certain threshold. Indeed, US commanders often go so far as to enumerate what loss of civilian life in a given attack would be justified as proportional to the direct military advantage to be gained. This can seem like a cold and morbid answer to an ethical question, but that is the degree to which US commanders go to clarify their intent and constrain their subordinates' use of violence. And they do so because they are ultimately responsible and accountable, legally and ethically, for the actions of those under their command.

This same framework of command and control would apply whether the subordinates in question are humans or machines. In both circumstances, human commanders are ethically and legally accountable for the use of violence, and it is incumbent upon them to understand the capabilities and limitations of their subordinates and to constrain their actions accordingly before giving them autonomy to use violence. If those subordinates happen to be intelligent machines instead of human beings, this does not mean there is no legal or ethical accountability for the use of violence. To the contrary, accountability for determining that someone or something is capable of performing specific tasks in war would still rest with human trainers and testers, while accountability for initiating an act of violence would still rest with the human commanders who knowingly ordered a person or machine to take that action. In both cases, humans remain accountable.

In some circumstances, commanders will want to give their subordinates greater freedom of action because they have determined that doing so is necessary to accomplish their mission or protect their troops or both. One way the US military addresses this challenge consistent with the laws of war is by declaring an "area of active hostilities." This is a limited geographic area in which commanders issue different rules of engagement that grant their subordinates greater autonomy to control the use of violence for a limited period of time. Commanders may permit their troops, for example, to presume that anyone or anything in this limited area is a combatant that can be fired upon without question. The main reason for doing so is because commanders have determined that achieving certain military objectives warrants less control over the use of violence. At the same time, declaring an area of active hostilities can limit accidental loss of life by signaling to civilians who may be in the area that they should relocate for their own safety.

Areas of active hostilities are an established military practice that human commanders could adapt to employ intelligent machines with greater autonomy to use violence in a future war. As today, an area of active hostilities would be an extreme circumstance that does not exist everywhere forever but only in a limited area for a limited time. And commanders might take this step for the same reasons they do today: to achieve their mission or protect the lives of their human troops or both.

Even with all of the laws and rules in place to govern the use of violence, militaries will continue to forbid, always and everywhere, certain actions as beyond the pale in the conduct of war. These are war crimes. Many of these actions, such as the intentional slaughter of civilians, will clearly remain war crimes, no matter what kind of weapons are used to commit them. But it is also possible that the concept of war crimes should be expanded in an era of intelligent machines. States may decide, for example, that it is an ethical imperative for human commanders to be able to communicate with their intelligent machines at all times because that ensures a human being is always in position to make the final decision about closing a kill chain and using violence in combat. For this reason, states might determine that destroying a military's ability to communicate with its own machines would constitute a war crime. Adapting long-standing legal and ethical concepts in this way would take work, but it would not be impossible.

But is it desirable? Would the prospect of removing more human beings from the physical conduct of war increase the temptation for states to go to war? No one should assume that a greater reliance on intelligent machines will make future wars bloodless. With the exception of satellites battling in outer space, future wars will likely still be fought in places where people are. And the enemy always gets a vote. Though a nation may want to remove its own military personnel from warfare, its enemies may retaliate with all manner of lethal weapons that threaten those military personnel as well as their civilian populations. I do not think that intelligent machines will fundamentally change that fact. And I am certainly not suggesting that America deny our military men and women access to more intelligent machines that could keep them safer and instead choose to put their lives at unnecessary risk, like hostages, in the hope that their mere human presence will make war less likely. That hardly seems ethical at all.

It speaks well of Americans that we put so much careful thought into these kinds of hard and important ethical questions. Our debates over the appropriate military uses of intelligent machines—over how to employ these and other new technologies consistent with our values, laws, policies, and established military practices—can be intense, but that is because so many Americans care so deeply about getting these hard questions right. I am not sure the same can be said about the Chinese Communist Party or Vladimir Putin. They are already using artificial intelligence to perfect authoritarianism and violate civil liberties on a massive scale within their countries. It is hard to imagine that the questions of human rights and dignity that correctly consume us are giving them equal pause in their rush to gain military advantage. And it is quite possible that their unscrupulousness could, in fact, enable them to get ahead.

The question for Americans is whether, or how, this changes things for us. That was the subject of a poll that the Brookings Institution released in August 2018.[6] When US respondents were asked what they thought about developing artificial intelligence and related technologies for warfare, 39 percent were opposed, 30 percent were supportive, and 32 percent were unsure. However, when those same Americans were asked whether our nation should develop such technologies for warfare if our rivals have done so, the percentage of those opposed dropped from 39 to 25 percent, and the percentage of those supportive rose from 30 to 45 percent. We cannot draw broad conclusions from one poll, but what many Americans likely realize is that great-power politics must shape our ethical decisions about the military uses of advanced technology. If we choose not to weaponize technologies such as artificial intelligence, that does not mean that our competitors will follow suit and be bound by the same choices. We do not have to be happy about this reality, but we cannot afford to deny it either.

We must also recognize an even more important reality: the values that will govern the future military uses of intelligent machines—and other emerging technologies—will be determined by the builders and users of those technologies. China and Russia will surely debate the finer points of international law with us as much as we want, but they will not stop trying to build more technologically advanced military forces. We can talk all we want about what our military and others should do and should not do with intelligent machines, but if we are not building these systems, and if our strategic competitors are, then we should not be surprised to find ourselves in the not-too-distant future living with much reduced influence in a world where lethal autonomous weapons are not only widespread but also routinely used for highly illiberal purposes—to enforce expansive and baseless territorial claims, curtail the independence of sovereign governments, oppress human rights, and ultimately threaten the American people.

I do not want to live in that world. I do not want to live in a world where the Chinese Communist Party is the only power with lethal autonomous weapons any more than I wanted to live in a world where the Soviet Union was the only power with nuclear weapons. The reason to build weapons is not because we want to but because we believe we have to, because we do not want to live disarmed and defenseless in a world full of predators. We should not build weapons because we are eager to use them but because we intend to make it so we never have to. We should build the most capable weapons we can because we want to prevent conflict and use of violence but also because, if that tragic day ever comes when peace breaks down and the men and women of America's military are sent to war, we do not want them going into harm's way with anything less than the best technology that our nation can provide. To do otherwise would only increase the risk that more of those Americans would not return home. That would be unethical.

That is why, if the question is whether the United States should build lethal autonomous weapons—and if we assume, as I think we must, that we will neither be able to trust nor verify that the Chinese Communist Party and other rivals are refraining from building such weapons themselves—I reluctantly say yes. But it is not an unconditional yes.

We should view lethal autonomous weapons in much the same way that we view nuclear weapons: we would neither want nor expect to use them, but we would make use of them every day to deter dangerous rivals from using such weapons against America and our allies. We should build these higher levels of autonomy into our intelligent machines to prevent hostile powers from utilizing similar capabilities to threaten our country and put our uniformed fellow citizens at greater risk. These weapons should exist for extreme cases of self-defense, and only then as options of last resort, similar to how we have viewed the fully autonomous settings on the Aegis and Patriot missile defense systems that we have used for decades. And just as those weapons have almost never been used, that should be the hope and expectation with any new lethal autonomous weapons: that level of autonomy on an intelligent machine should be a setting that we never intend to use. And if a person does choose to turn that setting on, human beings should be capable of turning it off.

One other principle should guide our development of lethal autonomous weapons, and it might make much of America's defense establishment deeply uncomfortable: radical transparency. This is largely the opposite of how the US government has approached the issue of armed drones over the past two decades. We went to enormous, perhaps even excessive, lengths to ensure human control over every task in the kill chain, but then we refused to talk about it. We treated the entire issue with the utmost secrecy—to our own detriment. We created the perception that we were doing something wrong and illegal, because we largely refused to explain, discuss, and defend our actions. We prevented the operators and overseers of these weapons from demonstrating how thoroughly and carefully they used these systems to limit unnecessary loss of innocent life and ensure that our actions were consistent with our values and our international legal obligations. The result was that America handed this entire ethical question over to our critics, who told lies about our use of military drones and made baseless accusations that we did not sufficiently and convincingly rebut.

That is a mistake we should not repeat. To the greatest extent possible, the US government and its partners in industry should be open with the country and the world about how we are grappling with the many difficult ethical and technological challenges associated with developing autonomous weapons in a responsible way that is consistent with US and international law. We should be equally open about the performance of these systems as we test and train them, and we should invite broad groups of engineers and ethicists alike to help make these machines more effective and predictable. Put simply, Americans should demonstrate how we are engineering our values into these machines. And we should challenge the Chinese Communist Party to be just as transparent about how it is addressing these ethical and technological questions as it builds its own weapons.

Perhaps the day will come when America and China will seriously negotiate limitations to our development, deployment, and potential use of lethal autonomous weapons, as is the hope of the many well-meaning negotiators meeting each year in Geneva. That is my hope, too, and it should be our goal.

But realistically, that day will only come, if it ever does, once both militaries have these weapons. Limiting the development and use of autonomous weapons will likely only be possible once the United States and its competitors can negotiate from positions of strength—with weapons to trade away—and are motivated by the fear that the continued construction of these weapons will endanger their security. That day has not yet come, and perhaps it never will, which is why America must focus now on building these technologies—or else we will find ourselves in a world in which a well-armed authoritarian great power writes the rules for advanced technology consistent with its values, not with ours.

EIGHT

A MILITARY INTERNET OF THINGS

In 2018, I traveled to Dayton, Ohio, home of the Wright Brothers, looking for the future of warfare. I had been searching for a while. One of McCain's first orders to me when he became chairman of the Senate Armed Services Committee in 2014 had been characteristically blunt as well as wise. "We've been investing way too much in the past," he told me. "I want to invest in the future."

The challenge we faced with respect to investing in the future reminded me of the old saying that the best time to plant a tree is twenty years ago and today. New military technologies and capabilities often take a long time to grow, and after years of military modernization paying the bill for current operations and over-budget procurement programs, one of the biggest challenges that McCain and I had was the ability to invest as much in the future as we wanted. Each year, I would help him identify billions of dollars of savings within the defense budget and reprioritize them toward new technologies. And we rarely got much help from the Department of Defense, which did shockingly little to highlight its most encouraging future-oriented programs. More often than not, it felt like my staff and I had to go searching for them, like Easter eggs.

My destination in Dayton was the Air Force Research Laboratory, where I went to learn more about an experimental, unmanned aircraft that was innocuously called the Low-Cost Attritable Aircraft Technology program. The basis for this aircraft was cheap target drones, which are designed to fly close to the speed of sound and maneuver like enemy aircraft so our fighter pilots can use them for target practice. Someone had the bright idea of turning this flying cannon fodder into a real military capability, and thus was born an aircraft called the XQ-58A.

The team in charge of the program explained that their goal was to develop an unmanned and increasingly autonomous plane—not a limited hobby drone, but a highly capable combat aircraft. The XQ-58A looks like the Air Force's most cutting-edge, high-performance fighter jets, and it would be able to fly more than twice as far. It would fly slower and carry fewer pounds of sensors and weapons, but it would fly at nearly the speed of sound. It could also be launched in place like a missile and recovered with a parachute, so it would not need runways on US air bases, which would likely be among the first casualties in any great-power war.[1]

Most importantly, the XQ-58A would be low cost, at least in Pentagon terms—so low cost that it would be what the laboratory team called "attritable." In other words, the Air Force could afford to buy a lot of them and lose a lot of them in combat. The XQ-58A is expected to cost several million dollars per aircraft once sensors and weapons are added, meaning that the Air Force could buy roughly a dozen or two XQ-58As for the price of one F-35A. Its first flight was delayed beyond my visit, but it took to the sky the next year with a new name: the Valkyrie.

A few months after my trip to Dayton, I flew to California to see another experimental autonomous system: the Navy's Extra-Large Unmanned Underwater Vehicle, or XLUUV. It had recently been pulled out of the water when I arrived, and I was able to walk around it and see it up close. Though far smaller than the Navy's manned, fast-attack submarines, the XLUUV lived up to its name. It is fifty-one feet long and capable of traveling sixty-five hundred nautical miles, which is roughly the distance from Los Angeles to Seoul, South Korea. The vehicle can grow an extra thirty-four feet with the addition of a payload module that, in the words of then deputy secretary of defense Robert Work, would enable it "to drop things out of the bottom or launch things out of the top."[2]

Here, too, one of the overriding virtues of the XLUUV was its cost. One of these systems costs roughly $55 million, and it would not be surprising if that cost doubled with the addition of the sensors and payloads that the Navy may include. That is far from cheap, but by comparison, the most capable variant of the Navy's Virginia-class submarine costs $3.2 billion per boat. It carries 120 sailors and takes roughly three years to build. The Virginia-class submarine is a far more capable platform in every way, but the amount of money that it costs for a single platform could buy roughly thirty XLUUVs, which also got a new name shortly after I saw it: the Orca.

Beyond the systems themselves, which were simply different kinds of military platforms, what excited me most about the Valkyrie and the Orca—and the reason I had traveled across the country to learn more about them—was the broader capability they could add up to: a large, distributed network of military systems that could be built and modernized faster, cheaper, more flexibly, and in greater numbers than any of our traditional military systems. These autonomous systems could also help to deliver even larger numbers of even smaller, lower-cost, but shorter-range autonomous systems to future battlefields far away from the United States. In short, I was less focused on what these systems were at the time than what they could become.

Both the Valkyrie and the Orca can operate with limited amounts of autonomy, performing basic operations safely and effectively with little human involvement. These are impressive technological feats, but these systems have so much more potential. To reach that potential requires the addition of artificial intelligence, vehicle autonomy, and other emerging technologies—mostly software more than hardware—that are being pioneered more by commercial technology companies than America's traditional defense industrial base. These technologies are what is needed to make unmanned systems such as these into intelligent machines.

When I say *intelligent machines,* I am not referring to current military drones, such as the Predator or Reaper aircraft, which are often viewed as the cutting edge of robotic warfare. In reality, these supposedly "unmanned" systems require dozens of people to pilot each one remotely, steer its sensors, maintain it on the ground, and analyze the information that it collects, much of which is discarded because there are simply not enough people to process all of it. Indeed, for years, the US military has supplied only a fraction of the drone missions that its commanders in combat have demanded. The problem has not been a lack of drones, but a lack of people. The technology did not exist to build military machines that did not depend on immense quantities of human labor to function—until now.

All of the pieces currently exist to build intelligent machines that can perform increasingly complex tasks. Lower-cost and highly capable sensors can enable machines to collect large amounts of data about their environment. Well-trained algorithms can sift through all of that data and identify the key pieces of information that humans have instructed the machines to find. Those algorithms can run right inside the machine itself, using the same kinds of edge computer and graphics processors that perform hundreds of trillions of operations per second and enable self-driving vehicles. Software-defined communications links can move information from machine to machine, even if they are not connected to a network. And improving robotics can enable machines to perform more and more complex physical tasks without direct human control. None of this is science fiction. The technology is here now.

What distinguishes an intelligent machine from any other is its ability to collect data, process it, extract mission-relevant information, interpret it to varying degrees of complexity, share it with any other military system and take actions based on that information without human beings controlling the system themselves. Some in the US military are clearly thinking about how to transform systems such as the Valkyrie and the Orca into more intelligent machines. But the prevailing view of the role that such a machine should play is often limited to serving as an auxiliary for a traditional manned platform—as an aerial refueling tanker, an advance scout or spy, or a "loyal wingman"—much like how the Navy viewed aircraft carriers solely as auxiliaries for battleships in the 1920s.

These are reasonable roles for intelligent machines initially, but they are a far cry from exploiting their full potential. In time, intelligent machines should not just enhance manned platforms; they should replace them. The real goal should be to build the next battle network around intelligent machines. And when it comes to building networks, the US military has a lot of catching up to do.

A battle network is the means by which militaries close the kill chain. It is what enables them to understand, decide, and act. Battle networks consist of people and things—things that sense, things that shoot, and things that share information. A battle network is often highly complex, but at the most basic level, it is only those three things. Sensors provide information about what is happening. That information is shared across the network to "shooters," which take action. That could mean the literal act of firing physical projectiles, but the US military often refers to *shooting* as a broader term that applies to enacting any decision, whether it be in the form of cyberattacks, electronic warfare, jamming communications, information warfare, or other means.

Each piece of the battle network is indispensable, but it is the sharing of information that is most important, and most often overlooked. Things that sense and shoot are interesting. Things that share information are not. They are unsexy. They do not star in action movies. Those who regard themselves as "warfighters" rarely like to be bothered with mundane technical issues such as information protocols and pathways. But they are what transform a mess of individual military platforms into one battle network, and in an increasingly automated military they will be far more important. Indeed, without the ability to share information well, there *is* no battle network, and the result is that it takes more time, more people, and more money to close the kill chain.

This concept of networking is one of the most under-appreciated but consequential ways that the commercial technology world has left the US military behind. Through painstaking, unglamorous, iterative work over the past two decades, commercial technology companies have hammered out the underlying information architectures that enable everything to connect to everything else. This is why we can run any application on any device on any network, and why we can freely change any of them at any time without any concern that important information or connections will be lost in the process. The central idea of this digital revolution, which has enabled the Internet of Things, is that individual platforms matter less than the broader network that they are part of.

This is not how US battle networks have been built. Many Americans think that their military closes the kill chain the way it appears in movies, with networks of dazzling machines stretching from beneath the oceans to beyond the heavens, all collecting information about our enemies, sharing it seamlessly and instantly across the network, and enabling humans in dimly lit rooms to understand events, make decisions, and direct actions against targets in real time. To be sure, some of our military systems can come pretty close to this. But the overall reality is rather different. If I had a dollar for every time US servicemembers have complained to me about the inability of their platforms and systems to share information with one another, I would be a rich man.

Systems that are capable of working together often do so in very linear and rigid ways. A specific sensor can share information with a specific shooter to close one kill chain, but do not try to substitute any part of that battle network. It is like a jigsaw puzzle that fits together only one way, which makes it unresponsive to changing events and more vulnerable to attack.

It is not even accurate to say that the United States has one joint battle network. The US military is more a collection of balkanized battle networks that require large quantities of time and human struggle to cohere. And that is the deeper problem with our current military business model: each sensor or shooter that we add to our battle networks to increase their speed and effectiveness requires corresponding additions of exponentially more money and manpower. And sooner or later, we will run out of both, especially in a long-term competition against China, which has four times as many people and could soon have an economy as big as ours.

Technology will never completely lift the fog of war, but the men and women in the US military deal with way more fog than necessary. Much of the information technology they use while on duty is many years behind the current state of the art. The problem is not that the US military is on the verge of taking humans "out of the loop" of the kill chain but that the US military today has way too many loops and way too many humans in the middle of all of them.

I have watched these Americans in action in US military operations centers all around the world. The walls of these rooms are usually covered with flat screens displaying live video from military drones or maps of the battlespace, and the floors are packed with desks where individual servicemembers monitor their narrow slice of the battle network on multiple computer monitors. In higher-level headquarters, operations centers can be the size of a basketball court, with more screens and nicer desks than the smaller, more spartan, tactical operations centers in war zones, which might have plywood desks and fewer flat screens. Operations centers, regardless of their size, are like the brains into which much of the information from military systems and sensors flows. In the absence of machines that can share information directly with other machines, this is how the United States connects its battle networks: a lot of people sitting in a lot of large rooms.

These people are some of the best men and women that America produces, and I have watched them spend excessive amounts of their time and talent trying to solve problems associated with closing the kill chain that better technology could solve for them right now. This starts with just making sense of events in the world—the first phase of the kill chain. Most military sensors are high-powered machines that require humans to operate them manually, like sound technicians on the sidelines at football games who point large receiver dishes at the players and try to pick up what they say on the field. For many years now, machine learning has been mastering all manner of narrow tasks, such as identifying people and objects in images. And yet, it is still overwhelmingly the job of human beings in the US military to make sense of the information each of their high-powered sensors collects and to figure out how to act on it.

While observing this, it occurred to me at one point that it is as if a separate person was in charge of operating and overseeing each of your five senses, and rather than having one neural network that can fuse all of this information together to generate understanding, what happens instead is that all of those people have to talk it out to try to make sense of the world. This might entail US servicemembers actually yelling back and forth across a large room. More often they use a computer-based instant messaging program called mIRC chat. I have watched individual servicemembers juggling a dozen separate chat windows, which can often involve taking information generated by one machine and manually transferring it to another machine. They call it "hand jamming" or "fat fingering." It is slow and prone to human error.

A friend of mine who recently did targeting in the US military told me that the best way his unit could get on one page in identifying a target was with Google Maps. They had to gather up all of their different streams of information about the target from their assorted sensor platforms, come to a time-consuming decision on where the target actually was located, and literally drop a pin in Google Maps to direct their shooters where on earth to fire their weapons. This was around the time that the Google employees wrote their open letter to their leadership demanding that the company cut ties with the Department of Defense lest their technology contribute to lethal military operations. "If those folks only knew how many bombs the US military has dropped using Google Maps," my friend told me, "their heads would explode."

Americans in uniform are often the first to joke about the inadequacies of many of the technologies they have to use, but I think it is a way of coping with their knowledge of the fact that these inadequacies could have deadly consequences. The US military fails to close the kill chain all the time. Threats appear, and Americans in uniform work frantically to understand them, decide what to do, and take actions—working far harder and longer than should be necessary because of outdated technologies, disjointed battle networks, and old ways of doing business—and while time passes, threats just vanish. The inability to close the kill chain could be of no consequence to those Americans. But it could mean that their unit walks into an ambush or that a missile hits their base or their ship. Through no fault of their own, through nothing they failed to do, it is these Americans who stand to lose the most.

The US military today is simply much slower and less effective than it could or should be at doing the one thing that will determine its success or failure—closing the kill chain. This problem has not been more apparent because we have mainly been fighting lesser opponents for three decades. We have been faster than they have been at closing the kill chain, even though we have been nowhere near as fast as we could or should be. This will not cut it in the future.

This is why we must view the emergence of intelligent machines not merely as a way to optimize our existing battle networks but also as an opportunity to break from our present model and reimagine it for the future. The new battle network should look less like our current military and more like the emerging Internet of Things, a network of intelligent machines that can collect, process, interpret, share, and act on information within the parameters of human-defined objectives. Each of those things can perform complex functions, such as regulating the temperature in our house, keeping watch over our front door, or serving as an in-home assistant for routine daily chores. But, ultimately, all of those things are just nodes in an ever-expanding Internet of Things that share information and facilitate understanding, decisions, and actions in our daily lives.

The emergence of intelligent machines will make it possible to build the future battle network as a kind of Military Internet of Things. One of those things might be an autonomous aircraft such as the Valkyrie, an unmanned submarine such as the Orca, or a small satellite such as the ones I saw at SpaceX. It could be a cyber payload. Or it could be a self-driving ground vehicle, ship, or amphibious system. What will matter far more than the things themselves is the connections between them—the ability of every sensor to share information with every shooter, every shooter to receive information from every sensor, and every machine to transmit information in real-time and at all times to every other machine. The things themselves are just nodes in the network—things that sense, shoot, and share information. The most important objective is for the battle network to facilitate human understanding, decisions, and actions.

A Military Internet of Things does not mean machines control the entire kill chain. Rather, the goal is to differentiate better between what intelligent machines can do and what human beings should do. Humans still do many things better than machines do, and this will likely remain the case for years or decades or possibly forever. Humans are superior to machines at putting pieces of information into broader contexts, inferring the meaning of events from the actions of objects, weighing the risks and trade-offs of different courses of action, and evaluating the strategic and ethical implications of actions. The purpose of a Military Internet of Things is not to replace people in the performance of these essential roles. To the contrary, it is to free up people in our military to focus more of their time on performing these core functions better.

A Military Internet of Things runs contrary to the platform-centered view of the world that prevails in much of America's defense establishment. Because we have always thought about building machines around human preferences and limitations, we tend to define our goals as platforms that are better, faster, and stronger than our current platforms. We have become so attached to the particular technologies that have delivered American military dominance for so long that we too often tend to mistake means for ends, desired things for desired outcomes.

However, technology itself is never the goal. It is always the means to achieving the goal. The real goal of sensing, for example, is not to collect exquisite sensors but rather to extend the reach and accuracy of human understanding. Similarly, the real goal of shooting is not to stockpile traditional arms but rather to extend the reach and efficacy of human action. Better platforms may be a means to an end. But the real objective is to have better, faster, more adaptable kill chains—to be able to understand, decide, and act more effectively under highly dynamic conditions than our opponents. The critical source of future military advantage will be the ability to impose so many complex dilemmas on our opponents at once that we shatter their kill chains, disrupt their ability to command and control their own forces, and leave them incapable of understanding what is happening, making sound decisions, and taking relevant actions.

That is what a Military Internet of Things will make possible. Ever since human beings have been building technology, one machine of any complexity has required many people to operate it. Often, most of those people act behind the scenes, but without them, the machine cannot function. This has meant that the growth of battle networks has always been limited by the availability of people. The basic math of battle networks has been that it takes many people to operate one machine of any complexity. This has been like the law of gravity—until now.

The next battle network, built around intelligent machines, will invert the ratio of humans to machines for the first time ever.

Instead of needing large numbers of people to operate one machine, one person will be able to command large groups of machines single-handedly. That has never happened before, and it will require a fundamentally more dynamic, flexible, and resilient approach to command and control that can pair sensors with shooters, decisions with actions, demands for military power with supplies of it countless times per day under any kind of operational scenario. This form of command and control could look less like what the US military practices today and more like the instantaneous operations of ride-sharing services such as Uber or Lyft.

A Military Internet of Things will be built upon the concept of human command and machine control. As machines become more intelligent and capable of performing basic tasks that have always before required humans to complete them, commanders will delegate more of those tasks to the emerging Military Internet of Things—the way civilians have begun to delegate monitoring events in their homes, navigating their way around the world, or even driving their cars to intelligent machines. This delegation will start with mundane functions such as moving machines or information from one place to another and identifying critical pieces of intelligence within oceans of data. But in time, humans will depend on machines to facilitate more and more tasks associated with understanding, decision making, and action.

Beyond involving fewer humans in command of more machines, a Military Internet of Things will also feature machines in command of other machines acting under the orders of humans. This can sound unsettling, but it already happens in rudimentary ways in our military and in more robust ways in our daily lives. In fact, intelligent machines that control other machines is the only way that the ever-expanding Internet of Things can function. A big, distributed group of machines can only grow so large before one of them has to be in charge of the others, lest the whole network devolve into chaos. Deciding which machine will lead is commonly done in the Internet of Things through a process of leader elections, when a routine program puts one machine in charge of others to ensure good order and discipline in the network, to direct the performance of basic tasks, and to ensure that machines are correctly following human orders.

If this sounds like a form of military command and control, that is because it is, although in the case of machines, each system is equally capable of becoming the leader at any time. If, for example, a group of machines gets cut off from the broader network, a leader election is run, and one of them takes charge. If the group is then rejoined to the rest of the network, the machines revert to their original hierarchy. This technology is the foundation upon which the commercial world has built increasingly larger networks of intelligent machines over the past decade, and most of that technology exists right now to build a new kind of battle network.

A Military Internet of Things will further alter the relationship between humans and machines because it will involve lower-ranking humans taking orders from machines that higher-ranking humans command. This may also sound unsettling in the abstract, but the same phenomenon is already functioning usefully in daily life. A similar dynamic happens, for instance, whenever we use apps such as Uber, Lyft, or InstaCart: people issue orders or decide to take actions; their commands go to intelligent machines, which interpret how, where, and when to perform those missions and then task people to do them; and those humans receive the orders and execute them, using less-intelligent machines to do so.

The reason that Uber drivers are not revolting over the fact that they spend their entire day literally being ordered around by intelligent machines is because they do not see it that way. They see it as following orders that a human has ultimately established for them, even if that human intent has been interpreted and refined by a machine. The same will hold true in military affairs. Senior commanders will issue orders to their subordinates, but those orders may be further specified by intelligent machines that have access to more and better real-time information. And a lower-ranking officer who receives those orders from a machine will view them no differently from how ride-share drivers view their daily taskings.

By inverting the ratio of humans to machines in the future battle network, a Military Internet of Things will arrest and reverse the shrinking of the US military that has consistently occurred for the past seventy-five years. The United States prevailed in World War II primarily by superior military mass. We simply outproduced our enemies. But in the early Cold War, the United States recognized that it could not win a numbers game against the Soviet Union, and we correctly made a bet in favor of quality over quantity. We used technology to field fewer numbers of more capable forces. And that is what we have been doing ever since.

As a result, the US military has certainly become more capable, but it has also gotten smaller and smaller—a trend that only accelerated amid the large military drawdown that occurred after the Cold War. For example, the US Air Force that fought in Iraq in 1991 had more than twice as many attack aircraft for operational missions as does the Air Force of today. Most of our current attack aircraft are unquestionably better, but as capable as any military system may be, it is not capable of being in multiple places at once.

A Military Internet of Things would reverse this trend completely. Because it will be built around intelligent machines, a future battle network could become exponentially larger than any military force ever assembled. For example, the US Air Force currently comprises 321,444 active duty personnel but fewer than 6,000 machines (aircraft, satellites, and nuclear forces). The US Navy currently has 326,046 active duty personnel but operates only 288 ships and submarines and 3,700 aircraft that are deployable. If each of those humans were instead commanding multiple intelligent machines, as will be possible with a Military Internet of Things, the next battle network would quickly grow to millions of systems. And that is before considering the Army and Marine Corps, which are primarily composed of people. This will be a return to a scale of battle networks that humanity has not seen for more than a century, or possibly ever.

The return of mass to military affairs will be possible because each intelligent machine will cost a fraction of the price of military machines today, which are built around humans. Putting people in machines makes them significantly more complex and expensive. A single F-35, for example, contains more than three hundred thousand separate parts.[3] Many of these parts, as in all machines that contain people, are required to ensure that their human occupants are safe, comfortable, and capable of controlling the machine. Most of those components will be unnecessary in intelligent machines, which drives down their complexity and cost and makes them easier and cheaper to produce, especially with advanced manufacturing methods.

In fact, many of the machines in a Military Internet of Things could become so inexpensive that they would not need to be maintained at all. They would be expendable, "attritable." This would enable the US military to acquire technology more as we do in the commercial world, where the emphasis is less on maintaining the same machines for decades at a time than on acquiring the latest technology as it becomes available. This is less possible with exquisite military systems that take many years to produce and have nine-digit or ten-digit price tags. But it is possible, even necessary, with lower-cost, "attritable" systems, where the goal is always to have the most cutting-edge capabilities—and lots of them.

It is not just the size of the battle network that will increase exponentially in a Military Internet of Things. So will its speed. The US military today moves information at a largely human pace. By contrast, a Military Internet of Things will share all information at all times, because all of its intelligent machines will effectively be part of the same brain. Rather than individual systems viewing the battlespace through their own narrow perspectives, there will be one network that sees the entire battlespace at all times, similar to how AlphaStar could see the entire game board while playing StarCraft II. When one machine identifies a relevant piece of information, it will be able to share it instantly across the network, regardless of where individual systems may be positioned. This will enable humans to close the kill chain at machine speeds. What now takes the US military minutes, hours, or even days will be done in seconds or less.

The speed and size of a Military Internet of Things could eventually increase to such an extent that its human operators struggle to keep up. They may find that they depend on intelligent machines not only to extend their own abilities to understand, decide, and act but also to manage the volume and velocity of military operations occurring at machine speeds. We must take such risks seriously and, rather than being cavalier with respect to the dangers of automation, build a command and control framework that safely integrates intelligent machines into our operations. But, although our growing dependence on technology and the Internet of Things is a legitimate concern—and one that many civilians struggle with in our daily lives—the larger concern in regard to the men and women of

America's military is not that they will become too dependent upon intelligent machines in the future but rather that their ability to succeed in the future is at risk now because of a lack of this technology.

In time, a Military Internet of Things could lead to a better division of labor between humans and machines. The essential function of any military machine is to facilitate human understanding, decision making, and action, which it enacts by sensing, shooting, and sharing information. But if a machine has to carry people or cater to human preferences and limitations, that becomes its most important mission, as it must, and the machine must be built completely differently. Aircraft, for example, require all kinds of additional complexity to support the life of the pilot, and their performance is limited by the human body's tolerance for gravitational forces in flight.

The same is true of sensors. Because it has mainly been a human task to operate sensors, most sensors are built around human preferences and limitations. Optical sensors, for example, have taken increasingly exquisite forms, such as high-resolution imagery and full-motion video, because that is what humans need to see things and understand the world. The problem is that these exquisite sensors often consume a lot of the precious resources that militaries will always find scarce, such as money, power, and network bandwidth.

Finding information within terabytes of sensor data, however, is one of the many narrow tasks that well-trained artificial intelligence can now perform far better, faster, and at vastly larger scales than people. And machines can do this work without many of those exquisite sensors that have been optimized for human use. This means that a Military Internet of Things will be built around the assumption that intelligent machines, not human beings, will be processing most of the information that the network collects. That leads to a set of design choices about how to build those machines that is totally different from the ones we use now. The emphasis will be more on quantity than quality, and the value of having a lot of sensors will be apparent to anyone who has ever played flashlight tag: The best way to find the other players as they run around and hide in the dark is not one brighter flashlight; it is more flashlights.

In time, as machines are built to function more autonomously, human preferences and limitations will be even less of a factor in how military machines will be designed. Those machines, even intelligent machines, are ultimately just trucks that carry the means to sense, shoot, and share information. But because they must regularly carry humans to make decisions for the machines, they must accommodate human preferences. But if the *only* purpose machines have to serve is to be trucks, rather than life-support systems and fighting platforms for human beings, these machines can be designed more simply and more in accordance with their primary purpose. Ships could come to resemble self-propelled missile barges, for example, and ground vehicles could come to resemble self-driving containers of rockets. This is not a new idea, but the emergence of intelligent machines will finally make it achievable in the next battle network.

A Military Internet of Things will not only refocus machines on their essential military purpose. It will also make the same thing possible for people. The primary purpose of human beings in military affairs is their moral agency, their capacity to make strategic, operational, and ethical decisions, especially about the use of violence in war. Just as machines are diverted from their primary purpose by having to cater to human preferences and limitations, so too are humans diverted from their primary purpose when they must do mundane, narrow tasks that require little or none of their moral agency—jobs that are better suited for intelligent machines.

This would be perhaps the most significant impact of a Military Internet of Things: it would liberate human beings from many of the repetitive, less valuable tasks that have traditionally consumed military commanders, and largely still do. People would have to worry less about the technical operations of the battle network—how specific sensors work together to generate understanding or how to move machines and information into the right places to take their desired actions. Instead, as machines become more intelligent and more capable of performing narrow military tasks, they could enable more people to focus on the more crucial military task of command—setting goals and missions, giving orders, ensuring that orders are followed, and making decisions that their subordinates cannot or should not make on their own.

This must be the ultimate objective—to focus machines on what they *can* do so that humans can focus on what they *must* do. This will take a long time to implement, but eventually, as machines become more intelligent and their numbers grow large enough, their users would cease to view them as things at all. They would care only about the services those things provide. This mentality is already unfolding with the Internet of Things. When people want to listen to music, they do not want to carry around particular albums anymore; they want to have access at all times to their entire library of music. When people order rides, they care less about who picks them up or what kind of vehicle they are driving and more about the ride arriving quickly and getting them to their destination effectively. The details of those services are left to an intelligent network of machines to implement those orders.

A Military Internet of Things would function the same way. The battle network would eventually contain so many intelligent machines that commanders would no longer think of them as aircraft, submarines, ground vehicles, or other specific platforms. They would pay less attention to whether those machines were operating at sea, on land, in the air, in space, or in cyberspace. Machines would not make such distinctions, and in time, neither would their human users. Those commanders would simply see intelligent machines everywhere, and the machines would eventually become so interchangeable, so expendable, and so ubiquitous that the things themselves would recede from view. Eventually, commanders would focus less on those traditional goods and more on the services that a Military Internet of Things provides—the ability to gain understanding, make decisions, and act. In short, human beings would be able to direct their undivided attention to the one aspect of warfare that is of greatest ethical and operational consequence: the kill chain.

NINE

MOVE, SHOOT, COMMUNICATE

Jan Bloch was not a soldier. He was a banker who was born into poverty in Warsaw in 1836 but worked his way up to become a wealthy railroad financier in Russian-controlled Poland. He never served a day of his life in uniform. But he was passionate about military issues and for years obsessively studied how the new technologies of his era would change warfare.

Bloch examined the introduction of the machine gun, smokeless gunpowder, long-range artillery, new types of explosives, railroads, telegraphs, steamships, and other innovations. And he traced their increasingly devastating effects from the Crimean War in the 1850s through the American Civil War a decade later, the Austro-Prussian War in 1866, the Franco-Prussian War in 1870, the Russo-Turkish War that began in 1877, and the start of the Boer Wars in 1880. He poured the results of his lifelong study of technology and warfare into a six-volume doorstop of a book that he published in 1898, four years before his death. He called it *The Future of War.*

What Bloch foresaw with stunning prescience was a future battlefield that would be far more lethal than most of his contemporaries imagined. The invention of smokeless gunpowder would literally lift the fog of war that had hung thick over past conflicts so that, unlike in previous skirmishes, opposing armies would remain dangerously exposed after the initial volleys of gunfire. Rifles could shoot farther, faster, and more accurately than ever. For centuries, the best professional soldiers could fire a few accurate shots per minute. At the end of the nineteenth century, average conscripts could fire dozens of accurate shots per second. And because bullets had become smaller, soldiers could carry more of them into combat. Modern fast-loading artillery, equipped with range finders and high-explosive shells, were 116 times deadlier, by Bloch's calculation, than guns from just a few decades prior.

What Bloch foresaw was a classic case of what Andrew Marshall later called a revolution in military affairs, when technological and other changes fundamentally alter how militaries build and operate their forces. For Bloch, this meant that battlefields would become killing fields, where combatants would never "get within one hundred yards of one another." War would cease to be "a hand-to-hand contest in which the combatants measure their physical and moral superiority." Instead, Bloch predicted, "the next war will be a great war of entrenchments."[1]

Although many of Bloch's predictions came to pass with eerie accuracy, he did get one big thing wrong, which is why history has largely forgotten him. Bloch believed that the sheer carnage of modern combat would be so appalling that large-scale war between the great powers would, as he put it, "become impossible." Of course, not a decade after Bloch's death, the nations of Europe marched into a great war in 1914 that unfolded largely as he had predicted. Forty million people were killed in four years.

Technology was not the only reason why the carnage of World War I was so horrific. It was also because militaries had radically changed what they fought with but not *how* they fought. Much of the war was waged with modern technology but antiquated doctrine.

Technology had increased the killing power of defenders in every way, but it had done little to advantage the offense, which still moved across the earth on foot, as the Roman legions had.[2] A generation of young men ended up in the trenches that Bloch had foreseen but were then sent repeatedly charging out of them, only to be annihilated by machine guns, modern artillery, and mustard gas.

Another factor that made the war so calamitous was the military technological parity that existed between the great powers. The existence of horrible new tools of death and destruction was made worse by the fact that nearly all of the combatants had nearly all of the same weapons. And they ground away at each other for years, struggling in their trenches to find an elusive advantage.

Similar dangers are developing today. The proliferation of information technologies and precision strike weapons over the past three decades, especially in China, has eroded America's longstanding military dominance. This has created significant advantages for defenders in the age-old military competition between offense and defense. The US military's ability to fight offensively still depends on the same kinds of large platforms that it has used for decades, but they are now more vulnerable than ever to the volumes of highly precise weapons that China has fielded in recent years. This dynamic is similar to what unfolded prior to World War I.

In another development reminiscent of great-power competition of the late nineteenth century, the United States and China will become peers that likely have the same technologies—those used to build more traditional arms as well as those enabling technologies that will transform the entire kill chain. One power may develop certain capabilities faster than the other, but they will generally compete in a state of military technological parity, where advantage is gained and lost less by the sheer fact of having a given technology than how fast it is acquired and how effectively it is put to use.

We must ask, then, how today's competitors, primarily the United States and China, will gain that advantage in the revolution in military affairs now under way. This, unfortunately, is unknowable. But military planners do not have the luxury of waiting to see how events will unfold. They must instead make consequential bets now about how to invest limited resources in wildly expensive things that often take a long time to arrive, and they must do it all on the basis of incomplete and imperfect understandings of what their competitors will do and how technology will develop in the future. Many of these bets will inevitably turn out to be wrong.

The best we can do is try to discern where and how we might gain future advantage in the enduring competitions that always have and always will define warfare. At the broadest level, there is the competition between offense and defense. More specifically, and most importantly, there is a competition over the kill chain—how militaries compete to gain better understanding, make better decisions, and take better actions, while denying these advantages to their rivals. But these competitions will develop amid more particular operational considerations that every new recruit learns in basic training: moving, shooting, and communicating.

It is not easy to peer into the future amid a revolution in military affairs and make high-stakes predictions about how it will turn out. But that is what Jan Bloch attempted to do in 1898. That is what Andrew Marshall sought to do in 1992. And that is what we must endeavor to do now.

Each of the three elements of warfare consists of a series of timeless competitions between combatants. Movement, for example, involves the competition of hiding versus seeking: attackers try to evade detection, and defenders try to find them. It also involves the competition between penetrating and repelling: attackers seek to force themselves into their opponents' spaces, while defenders seek to deny them access and drive them away. How militaries engage in these competitions has changed over time, but not the competitions themselves.

The United States has made several critical assumptions about movement in warfare over the past several decades. It has assumed that, in the event of a conflict, the US military would be able to move massive amounts of combat power thousands of miles from its domestic bases to the location of the fighting, and that it would have a lot of time to do so. It has further assumed that it could rely on relatively small numbers of technologically advanced systems to hide from enemy forces, penetrate into their territory, and overmatch them. It has assumed that the superior quality of its own forces could triumph over superior quantities of opposing forces. And it has assumed that it would be able to keep this war machine in motion with an unfettered flow of food, fuel, fresh equipment, ammunition, and other supplies from the homeland to the front lines.

China's military modernization has already called most of these assumptions into question, and emerging technologies will only make it harder for attackers to move. For starters, as sensors of all kinds become ubiquitous, including on future battlefields, hiding will be harder than ever and finding will be easier than ever, making it more difficult and riskier to penetrate another country's territory. The effects of this trend are already visible.

In 2014, for example, the Russian government emphatically denied what most of the world knew to be true: that Russia's Little Green Men had actively intervened in Ukraine. What revealed the truth (though Moscow never admitted it) was a flurry of pictures and videos of Russian forces and equipment that had been captured and shared on social media, including by Russian soldiers posing for selfies. This was also how it was revealed that Russia had supplied Ukrainian separatists with the surface-to-air missile system that shot down Malaysian Airlines Flight 17 on July 17, 2014. Civilians with smartphones captured the weapon moving away from the scene of the crime, which revealed its Russian military markings, and then they photographed the same system on its way back toward the border of Russia.[3] Similarly, when the Chinese government denied in 2016 that it was installing military capabilities on reclaimed islands in the South China Sea, commercial satellite imagery showed the truth in high resolution.[4]

Hiding from overhead surveillance will become infinitely harder as the heavens are filled with thousands of small satellites in the coming years. These large constellations of satellites will not only benefit consumers. As it becomes possible to blanket the earth with massive quantities of high-quality sensors—from cameras that take high-resolution pictures, to radars that track people and objects moving beneath them, to other means of knowing what is going on in the skies and on the ground with extreme precision and complete persistence—militaries will also benefit. They will no longer have to manage a scarcity of satellites, "revisiting" critical areas and hindered by long gaps in coverage, as has been the case since the dawn of the space age. Satellites will always be there, everywhere, providing constant surveillance of the entire world.

Militaries in the future will have little hope of hiding large traditional ships, aircraft, or ground force movements. Many technologies that enable militaries to hide may already be living on borrowed time. The German sensor manufacturer Hensoldt, for example, has claimed that its passive radar detected and tracked two F-35s from a pony farm on the outskirts of the Berlin Airshow in 2018.[5] Hiding from a single type of sensor is already becoming much more challenging, but what could be significantly more difficult is hiding from large networks of intelligent machines that can automatically fuse together many different types of sensors to provide combatants a clear understanding of the world. This could shift the advantage in the competition over movement from hiders to seekers and from attackers to defenders.

The balance could shift even further as more advanced sensors are fielded. Quantum sensors, for example, are being designed to detect the faintest disturbances in gravitational and magnetic fields that objects create as they move through the environment as if they are quantum shadows that have been forever hidden. Similarly, the US military is working on genetically engineering ocean plants so they can detect objects moving through the water by the chemicals, radiation, or other previously invisible signatures those objects emit.

Advanced sensors such as these are still many years off, but the trend is clear: hiding is becoming significantly harder, and militaries will need to search for new ways to conceal themselves beyond their traditional capabilities such as stealth. As sensors become ubiquitous, the best hope for hiding could be active steps to deceive the sensors themselves. This may become even more possible than it already is when militaries come to rely more on intelligent machines, rather than human beings, to process the information that sensors collect. The central front of the future competition over hiding and seeking could be the use of cyber and other digital tools to corrupt or trick the algorithms that militaries use to interpret their vast sums of sensor data. These small, fleeting blind spots could be the only opportunities left to hide.

Movement in the future will also entail a return of mass to the battlefield as militaries are able to deploy battle networks that resemble a Military Internet of Things. Fielding massive quantities of intelligent machines will enable more military systems to get in motion and stay in motion in more places than ever before. Militaries will face less of a trade-off between the quantity and quality of their forces. They will be able to have both.

Indeed, as it becomes more difficult to hide on future battlefields, militaries may even come to value hiding less. They could instead seek to be ubiquitous—to overwhelm their opponents with sheer numbers. A harbinger of this future was seen on September 14, 2019, when a rudimentary swarm of seventeen drones and eight cruise missiles of Iranian origin struck oil refineries in Saudi Arabia and knocked half of its production facilities offline. Its US-supplied defenses failed to respond effectively to this "attritable" massed attack. At present, these kinds of small systems can move only relatively short distances. But as their ranges extend, and as they become more intelligent and capable, these robotic systems will make life difficult for the small numbers of large platforms and bases on which the US military still heavily depends.

The race to build ever larger battle networks of lower-cost autonomous systems will be a critical area where militaries compete for future advantage. The sheer quantity of combat systems that militaries can field could determine whether the offense or the defense has the upper hand in operational exchanges. Mass will be increasingly essential to maneuver.

As the scale of movement on future battlefields grows exponentially, it will also increase the speed at which everything happens. Military power is inherently a scarce resource. But as the size of future battle networks grows, scarcity will give way to ubiquity. The time spent moving military systems from one place to another will collapse because the sheer quantity of systems will ensure that military power is available more often when and where commanders need it. The number of actions on the battlefield will increase, and the speeds at which they occur will accelerate in exponential terms.

Decisions that used to play out over hours or days will be compressed into seconds or less.

The speed of future military movement will accelerate further as more and more military things travel at hypersonic speeds, which is more than five times the speed of sound. Movement at these speeds will transform the timing and tempo of warfare. Vital national assets such as capital buildings, leadership headquarters, critical infrastructure, and command and control centers will all be just minutes away from hypersonic attack. Strategic decisions that affect global stability will shift onto tactical timelines that are reminiscent of gunfights between infantry units. This will necessitate increasing automation of self-defense, much like the Aegis missile defense systems provide, because people may not trust a "human in the loop" to respond quickly enough.

In reality, hypersonic movement will remain very expensive for the foreseeable future, so militaries will not have a lot of hypersonic weapons. They will likely hold them in reserve for the large forward land and sea bases of attacking forces, which will struggle to defend against these fast-moving weapons. Combatants will also struggle to defend the most valuable fixed targets in their homelands. As a result, they could shift to deterrence strategies to negate the offensive advantage of rival hypersonic weapons. What could emerge is a kind of mutually assured hypersonic destruction, where great powers defend high-value targets in their homelands by demonstrating that they can do similar damage to their rivals. This will catalyze the hypersonic arms race, but it could also discipline the "use it or lose it" fears that will emerge with these weapons. This is how we have avoided nuclear attack for decades.

The changing character of military movement will also extend to logistics—the ability to get forces on the move and keep them on the move. Logistics has been the greatest limiting factor in the history of warfare. Hence the old saying: "Amateurs talk tactics; professionals talk logistics."

The combination of intelligent machines and advanced manufacturing will transform how militaries keep their forces in motion. Each system will be easier, cheaper, and faster to produce than contemporary manned systems are. Advanced manufacturing will make production easier, cheaper, and faster still, and more importantly, it will enable militaries to shift their means of production closer to the battlefield. These future foundries will be able to print copies of weapons and machines, send them directly into combat, and quickly replace systems as they are lost in action.

As movement in the physical world becomes more congested and contested, militaries will prioritize movement in the digital world. Cyber has already become a domain of maneuver warfare, and this trend will accelerate as militaries become defined by digital technologies. The cyber domain provides military forces with greater opportunities to move faster, greater ability to hide from opponents, and greater flexibility to engage in effective acts of war that do not draw heightened scrutiny in the actual world. Because cyber capabilities will be essential to how militaries understand, decide, and act, they will focus on moving in the digital world.

The same dynamic applies to outer space. As military movement on Earth becomes harder, there will be a shift to the ultimate high ground off of Earth. Moving to and from space will become cheaper, easier, and more common, which means that militaries could eventually come to view space travel as little different from flying or sailing around the planet. Every phase of the kill chain will depend on military space systems. Militaries will establish bases on orbit to pre-position forces, manufacture reinforcements during a conflict, and deliver those machines right where they are needed on Earth in a matter of minutes. In the decades to come, movement on the future battlefield will extend to the space between Earth and the moon, transforming outer space into a new kind of highly contested battlefield.

Like movement, shooting also involves a timeless competition between combatants. It is the age-old fight between offensive and defensive fires: attackers seek to destroy rival forces and enable their own movement and communications, while defenders seek to shield themselves, shoot back, and deny attackers the ability to move and communicate. Shooting has consistently and considerably improved over time. But success is always a function of three factors that do not change: the range of fire (how far militaries can shoot), the accuracy of fire (how well they can hit what they are shooting at), and the effect of fire (how much damage they can do).

The US approach to shooting in recent decades has been shaped by its assumptions about movement. We have assumed that secure logistics and superior technology would enable US forces to get close to their targets and shoot a limited number of highly accurate shots. This has driven a preference for smaller rather than larger numbers of weapons and for shorter-range rather than longer-range weapons. We have also assumed that the best defense is effective forms of hiding—and that defending against enemy fires is less important because few of our opponents would be able to find US forces and close the kill chain against them.

The proliferation of longer-range, more accurate, and more lethal weapons is already overturning many of these assumptions, and it is disrupting the ability of US forces to move and communicate. This problem is most acute with respect to China, whose ability to shoot would make it difficult for US air and maritime forces to operate in East Asia during wartime. To a lesser extent, Russia is presenting the same problem to US air and ground forces in Europe. Even in the Middle East, which the US military has long considered a "permissive environment," Iran is fielding precision weapons, transferring them to proxies, such as the Houthi rebels in Yemen, and actually using them. Indeed, on June 20, 2019, Iranian forces shot down a US RQ-4 Global Hawk, a nearly $220 million surveillance drone,[6] with a precision surface-to-air missile.

The improved range, accuracy, and effects of fire have already created considerable advantages for defenders over attackers, and emerging technologies will likely further this trend. One main factor is the changing character of shooting. Bullets, bombs, and missiles will remain important, but as military machines become more intelligent, combatants will place greater value on "non-kinetic fires," such as cyber effects, electronic warfare, directed energy weapons, and communications jamming. The ability to corrupt the artificial intelligence at the core of a rival military or thoroughly disrupt the internal functioning of an opponent's government and society could cripple its ability and will to fight before it ever deploys forces.

Non-kinetic fires are one way that emerging technologies will increase the range of fire. What have traditionally limited the ability of militaries to shoot at longer ranges are the realities of physics, geography, and economics. It has been difficult to generate the power required to propel large objects over long distances, and it certainly has not been possible to do so cheaply. As a result, longer-range fires have traditionally been limited in number, relatively expensive, and used sparingly. This imposes physical limitations on the size of the battlespace, which has provided attackers places to hide and set up sanctuaries from which to mount offensive operations.

That will change. The laws of physics, geography, and economics apply little or not at all to cyber and other non-kinetic weapons. At the same time, the ability to shoot at longer ranges will increase as militaries field new kinds of fast-moving weapons that are hard to defend against, such as hypersonic weapons, supersonic cruise missiles, electromagnetic railguns, and cannons that fire hypervelocity projectiles over vast distances. These long-range weapons could make it more difficult for attacking forces to hide and move physically closer to their targets. This will expand the future battlefield to areas that the United States has traditionally viewed as sanctuaries, such as outer space, logistics networks, information and communications systems, and domestic critical infrastructure. Safe areas and sanctuaries will disappear. Everywhere will be contested and within range of enemy fires—even the US homeland.

The accuracy of fire has radically improved in recent decades, and it will only continue to do so, making the challenge of hiding in the physical battlespace even more difficult. Improved accuracy will have less to do with improvements in weapons than with the increased speeds at which intelligent machines can gather precise targeting data and share it with shooters. Even the US military has struggled to strike moving targets. This is largely because we have not had cohesive battle networks that could see into all of the dark spaces on the battlefield where militaries hide, find all of the moving targets there, share that information with other systems in real time, and close the kill chain. A Military Internet of Things could change this dynamic.

As militaries field intelligent machines by the thousands, they could illuminate more of those dark places, locate hard-to-find targets, and move that information to shooters at machine speeds. The result will be the prospect of real-time precision strike warfare under highly dynamic conditions. The goal of a Military Internet of Things is ubiquity—the ability of any sensor to enable any weapon to strike any kind of target at any time. Hiding on future battlefields will be hard enough, but once found, surviving will be just as hard. This could further exacerbate the attacker's current dilemma.

The growing accuracy of shooting will also contribute to increasing the effectiveness of shooting. Undoubtedly, militaries will be able to pack more destructive power into future weapons. But what will really increase the effects of future fires will be the significant increases in both the volume and the velocity of shooting that emerging technologies will make possible.

Nearly all of the intelligent machines in a future Military Internet of Things will be armed. They may carry bombs and missiles, but they will more likely be loaded with electronic attack and other non-kinetic weapons. In time, more of these systems will be armed with directed energy weapons that will enable them to shoot at the speed of light without the constraints of physical ammunition. The result will be an exponential increase in the number of available weapons such that when a future call for fires goes out, it will be far likelier that military systems are in the position to shoot in larger volumes and at greater velocities than ever.

Advanced manufacturing will also remove one of the biggest traditional limitations on the availability of weapons, which is the fact that they must be manufactured far away from the battlefield and then transported long distances to the places where they will actually be fired. As 3-D printing improves, however, military forces will be able to print more of their own ammunition near the battlefield. And the more weapons they have in the right places, the more willing and able they will be to fire them quickly. This will further increase the effect of fire.

In a broader sense, the real effect of future fires will be their ability to overwhelm targets with sheer mass. If a military were trying to defend against an approaching aircraft carrier, for example, it would not count on a few well-placed shots. The ship would have a better chance of defending itself and surviving against a small salvo of weapons. The more effective approach would be to try to overwhelm the aircraft carrier with thousands of intelligent machines, each of which could shoot many times on its own. Probably none of these attacks alone would be sufficient to sink a ship that large, but together they could destroy its aircraft, damage its control tower, crater its flight deck, and render it incapable of doing its job—what the military calls a "mission kill." This is the growing threat that all large bases and legacy platforms, not just aircraft carriers, face. And unfortunately, this logic is not lost on America's adversaries.

Perhaps the most important way that the character of war will change, more important even than moving and shooting, will pertain to communications. Like moving and shooting, military communications also consist of an enduring competition between combatants. It is the fight over information: militaries try to acquire critical information that enables them to close their kill chains while denying similar information to their enemies and preventing them from understanding, deciding, and acting. This competition is inextricably linked to the ability of commanders to control the flow of information to their forces, which enables them to move and shoot in timely, accurate ways. Communications are the links in any military's kill chain.

For decades, the United States has built military communications networks like bicycle wheels. The network is centralized in large hubs and military systems that depend on the hub for mission-critical information are connected to it like spokes. These hubs are large bases and operations centers full of people who are tasked with the tedious jobs of sifting through oceans of information, deciding what targets to take action against, and then directing military systems where to move and how to shoot. Collecting, computing, and transmitting all of this information requires enormous amounts of manpower, energy, network bandwidth, and physical space—all of which have only made US military communications more centralized, more fixed in place, slower to move information to those who need it, and a lot more vulnerable to attack.

In the future communications competition, militaries will find advantage in the ability to maintain the operations of their networks and the flow of information across them while jamming and attacking the networks of their rivals. The goal will be to build ever more resilient networks that can function securely, recover quickly, and reconstitute themselves even when under severe attack. Militaries will compete to build beyond-line-of-sight communications that function at greater ranges and software-defined communications that can hop across radio frequencies as attackers seek to shut them down. The new model of military communications will not be built around small numbers of centralized hubs but rather will push critical communications functions out to the edges of vast networks that are physically distributed, more secure, less vulnerable, and more resilient. Rather than bicycle wheels that can be shattered by taking out their hubs, the model will instead be centerless, reconfigurable mesh webs.

Decentralization of communications networks will accelerate as militaries become capable of their own version of "cutting the cord." Ubiquitous space-based communications networks will provide persistent access to information even in the most remote parts of the world. Regardless of where future military systems are on the planet—or, for that matter, off of it—they will always be in range of a satellite that can deliver the critical information they need. It will not be impossible to jam, disrupt, or destroy tens of thousands of small satellites distributed across low-earth orbit, but it could be prohibitively difficult and costly.

The decentralization of military communications will also be enabled by the emergence of intelligent machines. Future versions of systems such as the Valkyrie and the Orca, enhanced with edge computing and artificial intelligence, could make sense of the information that they collect, much as self-driving cars are increasingly capable of doing. Militaries will no longer have to move haystacks of data back to big operations centers where human analysts have to search for the needles within them. Instead, intelligent machines will be more able to find the needles themselves, and those small pieces of information are all they will have to move around the network. This is a fundamental change: the challenge that will be more important than communicating large quantities of data from machines to humans, as is done today, will be keeping humans in communication with large quantities of intelligent machines.

This points to the evolving role of traditional, manned military systems. The reason these systems were first created, and why they still exist, is to enable humans to move, shoot, and communicate. But as intelligent machines perform more military functions on their own, traditional manned systems will come to be seen more as mobile command and control centers. Their most important contribution will be their ability to keep human beings in communication with the numbers of intelligent machines under their command and executing their orders, not any particular sensor or weapon they carry. To the extent that future battlefield communications networks will still possess hubs, they will take the form of distributed manned systems acting as mobile command and control centers—not massive operations centers, but small teams of people in ground combat vehicles, onboard ships, or in aircraft.

This will be only a transitional role, however. Military communications will continue to become more decentralized and more distributed, and will require fewer people operating farther away from the immediate dangers of the actual battlefield. In time, the manned aircraft, ships, and vehicles that have been the backbone of militaries for a century or more will become less necessary even as means of communication, command, and control. There will be new ways for humans to oversee their battle networks and communicate with their growing numbers of intelligent machines. It might start with virtual or augmented reality and could progress to brain-computer interface—the ability for humans to send commands to intelligent machines using only their minds.

This seems far-fetched, but it already exists in limited forms. DARPA demonstrated in 2018 that it was possible for one person to control three drones using surgical implants that communicated that person's brain signals directly to the aircraft. What's more, the drones were able to send information they collected directly back to the person's brain, enabling the human user to perceive the drone's environment.[7] As biotechnology progresses, brain-computer interface may eventually enable human beings to view intelligent machines as extensions of themselves that enhance their abilities to understand events, make decisions, and take actions.

Greater fusion of human and machine intelligence, literally or figuratively, will provide militaries with critical advantages in future fights over information. In time, commanders will not make a single decision without assistance from a vast Military Internet of Things that they will communicate with constantly and directly. Intelligent machines will identify targets for human commanders and recommend decisions and courses of action about higher-level military strategy and operations. This may be the most important way militaries will accelerate their ability to close the kill chain: having machines perform the tasks machines do best and communicate information rapidly to their human commanders so they can make the operational and ethical decisions, as they must, especially regarding the use of violence.

The intelligence of machines could be the next major battlefield in the age-old fight over information. Humans will use intelligent machines for military purposes only if they trust them to perform missions safely, reliably, and effectively. Machine behavior depends on the integrity of the data that trains the machines' algorithms. If militaries are not confident that their intelligent machines will operate in combat the way they have performed in training, they might be compelled to sacrifice significant portions of their military advantage, lest they risk sending unpredictable, unsafe, and ineffective machines to perform military operations. For this reason, a military's data could become one of the most sought-after targets for its rivals, and nations will go to great pains to defend their data from adversarial attempts to attack or "poison" it.

As nations depend more on artificial intelligence for all manner of military functions, from target recognition to command and control, attackers will devise new ways to deceive intelligent machines. Researchers have shown how applying special stickers to images or objects can trick computer vision algorithms into drawing the wrong conclusions with great confidence. Some oft-cited experiments involve an algorithm mistaking a banana for a toaster and a self-driving car mistaking a stop sign for a speed limit sign. Militaries will undoubtedly seek to develop similar tools of deception to break the kill chains of their rivals.

How to counter these attacks and guard against the susceptibility of artificial intelligence to deception and disruption will be a central front of future competition. A priority will be accelerating the speed at which militaries can update their artificial intelligence. This will be a constant battle of move and countermove in which militaries try to stay one step ahead of their rivals, patching vulnerabilities and retraining algorithms with the latest information. The military that does this faster, the offense or the defense, will gain a highly fleeting advantage.

As militaries become more capable of deceiving or defeating algorithms that are interpreting a single kind of sensor, such as a radar or a camera, fusing many kinds of sensors together and rapidly contextualizing that information will become a critical source of military advantage. A future military algorithm, for example, may be fooled into thinking that it is looking at a banana instead of a tank. But if that object moves like a tank, sounds like a tank, has the heat signature of a tank, emits the electromagnetic signals of a tank, and exists in a place on Earth where tanks are likely to be, a more sophisticated sensor fusion algorithm may not be deceived. These kinds of contests between intelligent machines could determine who has the advantage in future fights over communications.

What does all this mean for America? In short, it means we have some big, big problems—but also some big opportunities if we reimagine our ways and means of warfare.

The United States has made major assumptions for decades about the character of moving, shooting, and communicating in future war. Many of these assumptions were reasonable at the time we made them, delivered overwhelming military advantages, and defined America's current force—a force that comprises lower numbers of technologically capable but highly expensive military platforms that have been optimized for moving relatively short distances away from their land or sea bases, hiding effectively, penetrating into enemy territory, communicating large quantities of information around the battlefield, shooting accurately but somewhat sparingly, and defeating larger numbers of less-capable forces.

Building the US military and operating it under these persistent assumptions has resulted in a long-running process of contraction, consolidation, and centralization around fewer numbers of platforms, bases, operations centers, communications networks, satellite constellations, and logistics forces.

The problem is that evolving threats and emerging technologies are calling into question these assumptions, and the sources of future military advantage will likely be built upon very different assumptions. Gaining that advantage will likely depend on succeeding in different kinds of military competitions. It will depend on finding those few and fleeting opportunities to hide. It will depend on fielding forces in ever larger numbers. It will depend on a scheme of maneuver that relies less on the traditional terrestrial domains of air, land, and sea and more on the non-terrestrial domains of cyber and space. It will depend on the speed with which militaries can replace the extraordinarily high losses of "attritable" machines and weapons that they will experience in combat. It will depend on new ways and means of self-defense and survival on battlefields swarming with accurate, longer-range, more effective weapons. And it will depend on the speed of communications in highly distributed networks and how quickly those networks can be reconstituted under attack.

The outcome of these and other competitions related to moving, shooting, and communicating—and at a deeper level, understanding, deciding, and acting—will determine future military advantage. In these competitions, however, success will not only depend on what we do but also, as always, what our competitors do. The United States should assume that China, in particular, is racing to gain advantage in the same competitions, with the same—and, at times, better—technologies, and that the result could be a future Chinese military that shares most if not all of the same core capabilities and characteristics as our military could have. If that happens, US military dominance will likely continue to erode. We will be living in a fundamentally different world than the one we have become accustomed to over the past three decades.

It is impossible to know which side, offense or defense, will benefit more from emerging technologies and the future of warfare. Defenders have the upper hand now, as in Bloch's era. But the character of war is always changing. Indeed, two decades after World War I, dramatic improvements in aircraft and ground vehicles created major offensive maneuver advantages, and Germany's *blitzkrieg* rolled over France's Maginot Line.

The United States must be attentive to how threats and technology will inevitably shift, especially after being ambushed by both so recently. But for now, as the proliferation of revolutionary new technologies and the emergence of a peer competitor more powerful than any America has ever faced contribute to a steady erosion of US military dominance, we face a far more immediate challenge: how to reimagine America's national defense in the absence of dominance.

TEN

DEFENSE WITHOUT DOMINANCE

On October 27, 2017, I helped John McCain write and send a letter to Secretary of Defense James Mattis. It has never been publicly released. The topic was the *National Defense Strategy*, which Mattis was drafting to meet the requirements of a law that I had assisted McCain in writing and passing the prior year. The letter was a plea to Mattis, after a decade and a half in which the US military had been consumed by counterterrorism operations, to shift the focus of our national defense and "prioritize the challenges presented by Russia and China."

"We no longer enjoy the wide margins of power we once had," McCain wrote, because America's military advantage had "declined precipitously" as great-power competitors, primarily China, were modernizing their forces and eroding America's military dominance. "We cannot do everything we want everywhere," McCain wrote. "We must choose. We must prioritize." And though money was vital, we could not "'buy our way out' of our current predicament." We had to think differently, and time was running out. The new defense strategy, McCain wrote Mattis, was "perhaps the last opportunity to develop an effective approach" to China before it was too late.

The reason we advocated for concentrating on China was not because we believed that conflict is inevitable or desirable. Far from it. Rather, it was because China is America's most capable military competitor, and it could present the US military with the most stressing operational problems. The United States had to face that threat honestly and build its future military purposefully to be capable of defending against it—if, God forbid, it was ever called to do so. Not only would that level of military strength help to maintain peace and deter conflict with an increasingly powerful China, it would also mean that the US military would be capable of dealing with all of the lesser (but still real and serious) threats posed by adversaries such as Russia, Iran, or North Korea. That is what McCain wanted the new defense strategy to prioritize.

Strategy is perhaps the most abused word in Washington. US leaders use it regularly to make all manner of lesser priorities sound like greater ones. That used to drive McCain crazy. "If everything is important," he would say, "nothing is important." Government strategies are more often laundry lists of hopes and dreams that help leaders avoid making choices. They seek to be inclusive of everyone's priorities and give every kid a trophy, rather than picking winners and losers among priorities that are all competing for finite resources. They say everything—and thus, nothing.

This was a risk the *National Defense Strategy* was facing as 2017 wore on. My staff and I had met regularly with Mattis's staff, who were helping him prepare the strategy document. We had known each other for years and shared many of the same goals for the new strategy. By the fall, the time for making choices with real trade-offs had arrived, and Mattis's staff were facing growing internal opposition, including from some of his own military advisers. Mattis was consumed with a full plate of crises in Washington and the world, and the signature achievements of the strategy were at risk of being cut out almost completely. I relayed all of this to McCain, who listened, smiled, and said: "Let's write the secretary a letter."

That letter would be one of the last times McCain left his mark on the new defense strategy, but it was hardly the first. Throughout 2017, I had helped him push his vision of change for US national defense through a litany of public hearings, classified briefings, statements, speeches, letters, draft legislation, meetings with senior defense and military officials, overseas travel, and even a thirty-three-page defense policy paper that McCain released on the eve of Donald Trump's inauguration. "If all we do is buy more of the same," McCain wrote in that paper, "it is not only a bad investment; it is dangerous."

This followed years of even more sweeping legislative activity, when we sought to change how the Department of Defense developed and bought military systems, break down barriers that prevented the military from getting access to more advanced technologies, eliminate wasteful or unnecessary spending to free up money for real military capabilities, boost defense investments after years of erratic and deficient funding, and shift as much of that money as possible from outdated and less effective programs to new, future-oriented priorities. And what motivated it all was our growing concern that America was losing its military advantage to great-power competitors, most of all China.

This is why we created the requirement of the *National Defense Strategy* and agitated for it to come out right—not because a paper strategy is an end in itself, but because, if done well, it can provide a point of reference for leaders who have to make hard choices between competing priorities and finite resources. This is largely what emerged in the *National Defense Strategy of 2018*, which Mattis released three months after receiving McCain's letter. It was not perfect. But it mostly got the big things right. It clearly defined the top priority of the Department of Defense as "long-term strategic competitions with China and Russia."[1] And it provided adequate detail in classified form, as required by the law, to guide the kinds of real policy and programmatic choices that McCain wanted.

The *National Defense Strategy* can best be thought of as an opportunity—perhaps the best opportunity in two decades—to redefine America's national defense. And the response thus far, especially since McCain's death, has been rather encouraging. Pentagon leaders have embraced the strategy in a way that has seldom happened before. Many senior military chiefs are beginning to rethink how, and with what, their forces will need to fight in the future. And leading defense thinkers are doing essential work to challenge and improve upon the strategy.[2] In short, a lot of people are saying a lot of the right things. But the main problem in recent decades has not been a failure to say the right things. It has been a failure to do enough of those right things.

America cannot afford to be ambushed by the future again, because the consequences this time would be an inability to defend not only the people, places, and things in the world that matter most to Americans but also the US homeland itself. America's strategic margin for error has disappeared. The *National Defense Strategy* provides a good start to a better set of answers, but it does not go far enough. Because we have put off change for so long, the scale of change now required is more extreme. New technologies alone will not save us. We need new thinking—an ambitious effort to reimagine the ends, ways, and means of US military power, as well as the role of our allies in this effort—to succeed in a future world where America's military superiority will likely erode further if China's military technological development continues.

In short, we need a strategy of defense without dominance.

A new defense strategy must start by rethinking America's goals. The idea of refocusing US national security on great-power competition, especially with China, has become a form of conventional wisdom in Washington in just a few years. The US military is only one part—and arguably not the most important part—of an effective response to that goal. The real question is what the United States is trying to *achieve* in this new era of great-power competition, especially militarily. After all, competing is not an end in itself. The bigger challenge is that the return of great-power competition and the relative decline of US military dominance require us to think differently about our goals from how we have grown accustomed.

Since the end of the Cold War, US leaders have gotten used to defining America's goals in the world rather expansively. No other great power threatened America's position of global primacy, so we were able to focus less on preventing bad things and more on enabling good things. This bipartisan desire to sustain American dominance and improve the world was behind many of the wide-ranging goals that US leaders defined for their military, such as humanitarian intervention, regime change, nation building, and broadly fostering a "rules-based" or "liberal" world order.

We defined our defense and military objectives without a great deal of consideration for external limitations on our ambitions, because they scarcely existed. America was dominant, and we largely called the shots on the international security issues that mattered most. The only meaningful constraints on America's goals were ones that we imposed on ourselves, and we did not always impose many. If US leaders chose to liberate Kuwait, stop ethnic cleansing in the Balkans, remove Saddam Hussein from power, or depose Muammar Qaddafi in Libya, there was little doubt that the US military could do so, and that no foreign power could stop us.

It is difficult to overstate what a complete anomaly the past three decades have been in the broad sweep of world history. This era of unrivaled American dominance stands in marked contrast to the rest of history, which has always been characterized instead by great-power competitions. And one of the defining realities of those competitions is that great powers are willing and able to impose real limits on each other's ambitions, especially with military power.

This is the world that the United States has now reentered, especially in relation to China, which is on pace to become far more than a great power. To be sure, the country has a host of internal problems that could hobble its continued growth. But if China does continue to gain wealth and power, it will become a peer that could achieve economic, technological, and military parity with the United States—and whose capabilities, in some instances, may even exceed ours.

If this happens, no operational or technological wizardry will enable America to roll back the clock to a time when we could do nearly anything we wanted unchallenged. Instead, we will need to relearn a lesson of history that we largely forgot during our three decades of uncontested dominance: that great powers are capable of limiting one another's ambitions and rendering many of each other's goals impractical or unachievable, regardless of how desirable those goals may be for one side or the other. Great powers force each other to define their core interests, the things each is truly willing to fight over, and then make compromises and accommodations as necessary over the rest, lest competition descend into conflict. This is the messy, unsatisfying, and oft-neglected other side of great-power competition. Call it the management of strategic rivalry.

This is already the reality with China. It is unlikely, for example, that a US president would send an aircraft carrier through the

Taiwan Strait in a significant crisis with China the way President Bill Clinton did in 1996. US carriers would probably not even operate within a thousand miles of the Chinese coast in the event of a conflict. Similarly, after Russian forces intervened in Syria, even leaders such as McCain who had pushed for more expansive goals for US policy began to concede that those goals were no longer viable because of the increased military and political risks they entailed. Predicaments such as these have less to do with what America wants and more to do with what a great power can deny or deter America from achieving.

With regard to China, the United States may not be able to recover the position of military dominance that we have long enjoyed and the abundance of security that it provided. But we are capable of achieving a goal that can defend the core interests of the American people, albeit a less expansive goal than our world-ordering ambitions of recent decades. China may be capable of denying dominance to America, but America can do the same to China. And that should be our goal: preventing China from achieving a position of military dominance in Asia, which might be accompanied by a growing global assertiveness that could lead to even more detrimental consequences for the United States and our closest allies.

This dangerous position is exactly where things are headed. If the United States continues with business as usual, it could soon find itself in a world where the Chinese Communist Party is capable of imposing its will militarily on any actor in Asia, including the United States and its allies, and exerting control over the center of the global economy on which the jobs, livelihoods, and security of Americans depend. This would give Chinese rulers the ultimate leverage in any dispute, military or otherwise. They would know that they could get their way through force or coercion on issue after issue of critical importance to the American people, that they could drive wedges between the United States and our allies, and that there would be little that the US government could do about it. This is how hard power and soft power are inherently linked.

How we think about the defensive goal of denying military dominance to China is constrained by another old reality of great-power competition: the prospect that a future conventional conflict with a great power, especially a peer competitor such as China with a technologically advanced military on par with ours, could extend to the US homeland. This is a reality that most Americans and their military are wholly unprepared for. Although it is true that the United States has lived for decades with foreign nuclear arsenals pointed at our cities, we have deterred that threat with our own nuclear weapons and a stated willingness to use them. What we have not considered, however, is that a foreign competitor would be willing or able to target the continental United States with large numbers of *conventional* weapons.

Indeed, the thought of the US military fighting to defend its homeland is a foreign concept for most Americans, and for our military. It is seen as something that the United States forces others to do but does not have to do itself. To be sure, homeland defense has always been the stated top priority of the Department of Defense, but the US military's role in that mission has been confined to limited missile defense, support for domestic law enforcement, and confronting US enemies far from our shores. US leaders never saw a need for real military defenses at home, so we did not build them. We optimized the US military to project power overseas. Surrounded by two oceans and harboring the largest and most powerful military in the world, we were not unreasonable in making this assumption, but as other nations have built long-range precision strike weapons and power projection capabilities of their own with an eye toward challenging America's defense, we have left most US territory vulnerable to conventional military attacks.

That is exactly the kind of threat that emerging technologies will make more possible. Great-power competitors know where to find the most important targets in US territory, such as national and military command and control installations and other critical infrastructure, and they are improving their ability to strike them at longer ranges, more precisely, and with greater effect. A future conflict could involve persistent cyberattacks, advanced cruise missiles, hypersonic weapons, and other more intelligent machines all striking targets in the United States. They could be launched from the territories of our rivals or from submarines and aircraft that could slip in close to US territory. As a result, for the first time since the nineteenth century, real homeland defense will have to become an American goal that consumes far more of our defense budget.

China's growing ability to strike the US homeland with conventional weapons may deter us from pursuing certain military objectives in a future conflict. When a peer military can hit targets on US territory with similar, or even better, weapons than what we possess, it puts the prospect of initiating offensive strikes against that nation's territory in a new and more visceral light. US leaders would still be free to act, but that action might involve direct, retaliatory consequences against our homeland that would make war real for Americans in ways that they have never contemplated. What is more likely is that the United States and China would seek to mutually deter each other and rule out each other's homelands as legitimate targets in a conventional war. That is a new idea for American political leaders and the US military, which have long assumed they could take war onto the soil of any adversary with little to no direct military consequences for the American people at home. The growing impracticality of this approach with regard to China will limit US military goals in the future.

To deny China military dominance, Americans must recognize that this is not just another defense priority among others—it must be the defense priority to which all others are subordinated. This does not mean a new cold war is upon us. That is not the reality of the competition with China. Asia is not divided into rival territorial blocs the way Europe was during the Cold War, and Asian countries will rightly resist being forced to pick sides between America and China. What this does mean, though, is that for the US military to deny military dominance to China, as it must, it will have to do less everywhere else.

In recent decades, US leaders have given our military too many missions and have prioritized US military "presence" in too many places across the world that deliver too little benefit to our national defense. American leaders must tell our military what it no longer has to do. And they will certainly need to avoid saddling our military with costly and unnecessary new missions, such as a war with Iran, an intervention in Venezuela, or preemptive military action against North Korea. Certain other missions, such as limited counterterrorism operations, can and should continue, and US leaders must realize that inattention to certain problems might allow them to metastasize into larger, more demanding threats. But overall, the US military will have to do less. This will require hard choices with real consequences, but we must make them.

Put simply, conserving US strategic resources—not just our military power but also our money, our leaders' time, and our allies' goodwill, among other things—must become a goal of US defense strategy. It will require US leaders to settle for less desirable, less optimal, and even riskier outcomes on other foreign policy goals so that we can prioritize the more important goal of balancing Chinese military power and harboring our own. This is why, for example,

Trump's decision to withdraw from the nuclear agreement with Iran was a mistake—not because the deal was "good," but rather because it would have enabled America to spend less of its limited military power focused on what is ultimately a secondary priority. US leaders do not need to be happy about this kind of decision making, but they should be prepared to do a lot more of it.

The United States cannot and should not contest every difference that it has with China militarily. To the contrary, US leaders will have to determine what our nation's core interests really are. After years of suggesting that the United States is prepared to fight for all manner of expansive and unclear objectives, we will have to decide what we are really prepared to fight for, especially in light of the scale of destruction that would result from a future war against a peer competitor with military capabilities that equal and possibly exceed ours. Certain core interests are still worth fighting for, but we must pare down that list to the essentials.

It is not just what America is prepared to fight for that must change but also *how* the US military plans to fight. This should flow directly from the goal of denying military dominance to China—a defensive objective, not an offensive one. It is less about the good things that military power could make possible than the bad things that we need it to prevent. As a result, the United States must change the way that its military operates and devise what Chris Dougherty, a defense expert who helped to draft the *National Defense Strategy*, has called a "new American way of war."[3] US thinking about warfare must shift from an offensive to a defensive mind-set. In short, America needs to put the "defense" back in our defense strategy.

This would be a fundamental shift. Since the end of the Cold War, and perhaps as far back as World War II, the American way of war has been fundamentally offensive. That is not a comment on *why* we fought wars but *how* we fought them. We have fought offensively. We have sought to project massive amounts of combat power far away from home, penetrate deep into enemy territory, use advanced technologies to evade and dominate opponents, take over and occupy their physical space, and stay there for as long as Washington has wanted. And there has been little, if anything, that America's opponents could do about it. Even when others have initiated hostilities, the US plan has always been to quickly go on the offensive, as we did after Iraq invaded Kuwait in 1990 and against the Taliban and Al-Qaeda after September 11, 2001.

The bad news is that the way the United States has been planning to operate its military for decades and at great cost—projecting power and fighting offensively—has become extremely difficult with regard to China and will become even more so. The United States already finds itself on the horns of a dilemma because of China's development of massive arsenals of advanced precision strike weapons that can find and attack the large bases and platforms that enable the United States to project military power. The problem is not unsolvable; there is much the US military could do to address it more effectively now with available technologies. That said, this problem will become more acute if China continues to move urgently to use emerging technologies to strengthen its military. If America remains attached to its traditional, offensive way of war, new technology alone will not save us.

What this means is that we are headed back to the future. If the United States and China make similar assumptions and use emerging technologies to build similar kinds of military capabilities, especially massive battle networks of intelligent machines, a great-power war waged by these technologically advanced competitors would likely be governed by the brutal, unforgiving logic of World War I:

forces that are entrenched in defensive positions could stand a decent chance of surviving and fighting effectively, but the moment they step off from their points of departure and try to advance against their opponents, they would likely enter a new "no man's land" that is teeming with ubiquitous sensors, intelligent machines, and advanced weapons, operating from the ocean floor to outer space, that are capable of closing the kill chain at scales and speeds that attacking forces would struggle to survive. And like World War I, conflicts between peer competitors fighting with most, if not all, of the same weapons would likely erode into stalemate.

This is not all bad news for America. Projecting military power and fighting offensively are not becoming more difficult only for us but *for everyone*, including China, which is also making considerable investments in large ships and other traditional platforms for the purposes of projecting conventional military power. If the United States develops a new, defensive way of war that is focused less on projecting military power than on countering the ability of others to do so, we could create the same dilemmas for our competitors that we are facing. In this way, emerging technologies could be tailwinds rather than headwinds. We could achieve the more limited, defensive goal of denying military dominance to China by creating the same kinds of anti-access and area denial predicaments for China that it has been creating for us.

This way of warfare would be much less about maneuvering offensively, penetrating into an opponent's space, attacking, and dominating an adversary in its own territory—all of which could be impractical against a peer competitor. It would instead be, more modestly, about denying the Chinese Communist Party the ability to impose its will militarily in US territory and on the people, places, and things that matter most to America in the world. The purpose would be to deter acts of aggression and war by demonstrating to potential aggressors that the US military can destroy any forces they send on the offensive, prevent opponents from projecting military power beyond their own territory, replenish our losses faster and more cheaply than they can, sustain the fight for as long as necessary, and halt their ability to keep attacking. It would be an American version of "winning without fighting."

This will not be possible, however, if the United States continues to plan to mobilize for war the way it has for decades, where most of the force has to flow from US bases to forward positions over weeks and months before it is ready to fight. That force would be attacked as soon as it started mobilizing in the United States. Its logistical operations would be hacked and fired upon. Its communications would be jammed. Its satellites would be disrupted and shut down. It would be attacked each step of the way on its journey across the world to the battlefield. And if that force actually got where it was needed, it would likely arrive too late to matter.

Indeed, that is exactly how China plans to win a future war in Asia and how Russia plans to prevail in Europe: strike rapidly, consolidate their gains before US forces can respond effectively, harden their victory into a fait accompli, and force the United States to escalate the conflict to attack and dislodge their forces. This kind of rapid aggression will only become easier when future war is moving at the speed of hypersonic weapons and intelligent machines.

To deter this kind of conflict, the United States must have nearly all of the military forces required to defend against great power aggression right where war might occur. This necessitates positioning large numbers of new military forces, especially autonomous systems, advanced missiles, and electronic attack systems, in Europe and Asia. It will also require the eventual forward deployment of advanced manufacturing and other means of production that could rapidly generate vast quantities of replacement forces in the event of conflict, where losses would be significant. If the US military does not plan to fight this way, it will find itself increasingly irrelevant to the future of warfare and at greater risk of failing to deter a future conflict—or even losing a war by failing to show up in time to defend effectively.

The purpose of preparing for war in this way is to never have to fight one. But even if the United States reestablishes its ability to deter conventional conflict, that will not mean an end to military competition or even conflict. To the contrary, competition will continue to shift more aggressively into the large and growing "gray zone" between war and peace.[4] These new shadow wars are already being fought in cyberspace, in outer space, in the public information domain, and between opposing forces of Little Green Men and other clandestine operatives. Our competitors, especially China, are fighting these kinds of asymmetric battles against the United States because they have been deterred from challenging us through more traditional ways and means of warfare. The problem is that our ability to deter conventional war is deteriorating. If we reimagine our defense strategy and restore conventional deterrence, the price of success will be fighting increasingly pitched battles in the gray zone. However, if China comes to believe that it could defeat America in a conventional war, it could embolden the Chinese Communist Party to confront us more directly. Then the gray zone would be the least of our problems.

Changing how and for what the United States would fight in the future is necessary but not sufficient. We also need to change what our military fights *with*. How the United States has built its military for decades is a direct result of the assumptions we have made about how we would fight if called upon to do so. We have assumed that warfare would consist of long, slow mobilizations to project massive amounts of military power across the world to far-flung battlefields where US forces could use superior technology to dominate weaker opponents while sustaining very few combat losses in the process. As a result, the US military today consists of relatively small numbers of rather large, exquisite, highly expensive, heavily manned, and hard-to-replace things.

What's more, all of these combat systems depend on an enterprise of communications centers, information networks, satellite constellations, and logistics forces, such as cargo ships and aerial refueling tankers that are not only similarly large, expensive, heavily manned, and hard to replace but also nearly defenseless, because they were built under the assumption that no US adversary would ever be able to attack them. Washington spends more than $730 billion each year on national defense, and this is what most of that money goes toward—developing, procuring, operating, maintaining, and crewing these kinds of traditional military systems.

The problem is that these small numbers of multi-billion-dollar systems are less likely to survive against the large quantities of precise, multi-million-dollar weapons that China, in particular, already has acquired—weapons that have been purpose-built to undermine how the US military plans to fight and to break its ability to close the kill chain. Some in Washington will wink and nod and say that there are things that can be done, like magic, to forestall this challenge. Some will also say that new technologies, such as artificial intelligence, can help keep America's legacy military effective for longer into the future. This is partly true. It is possible to buy time. But playing a losing game eventually ends in defeat.

The United States needs to build a different kind of military. And we cannot afford to repeat recent mistakes. Our focus must be on building and buying integrated networks of kill chains, not individual platforms and systems. We need to buy outcomes, not things. Those military things will matter less than the broader battle network that they add up to and its ability to facilitate human understanding, decision, and action. If we think about the problem this way, we can ask the right questions: How would we build the US military differently? What attributes should it have?

First, rather than small numbers of larger systems, the future force should be built around larger numbers of smaller systems. This will enable the US military to distribute more forces over broader areas. Our rivals would no longer be able to concentrate their sensors and shooters on a few big targets. Instead, they would have to find and attack many things over larger spaces. In this way, the United States could impose costs on its competitors rather than allowing them to impose costs on us, as is the present case. Every dollar that a competitor has to spend on more sensors and more weapons to target larger masses of US military systems is another dollar that is unavailable for them to spend on new offensive capabilities. This would be a reversal of the same kind of losing game the United States has been forced to play for too long.

Similarly, rather than expensive systems that are effectively irreplaceable, the future force should be built around lower-cost systems that are effectively expendable. If US systems are cheap to build, operate, and replenish, we would be more willing and able to lose them. This approach would also impose costs on our competitors. If it costs them more money and time to destroy our systems than it does for us to field and replace our systems, we could force our rivals to play another losing game: they can attack our cheaper systems with their more expensive weapons, which becomes unsustainable over time—or they can leave our military systems free to operate, enabling them to pose a continued threat.

This certainly argues in favor of new capabilities, such as networks of low-cost drones. But it also argues in favor of missiles—lots of missiles—many of which are available now but too often lose out in fights over funding to those large, expensive vehicles that members of the US military like to operate and members of Congress like to have built by their constituents. If the US Army and Marine Corps, for example, had more land-based missiles that could shoot enemy warships, it is true that the US Navy might need fewer warships, but it also means that the US military could more effectively defend its interests at sea. America can learn a thing or two in this regard from China, which has been aggressively expanding its missile arsenals for decades.

This points to another attribute that the future US military should have: rather than large numbers of people operating small numbers of heavily manned machines, the future force should consist of smaller numbers of people operating much larger numbers of highly intelligent unmanned machines. People are expensive. Putting people in machines is even more expensive. And no one ever wants to pay the ultimate price of losing a human life. Manned systems will not fare well on future battlefields, which will be extremely violent with heavy losses on all sides. Lower-cost intelligent machines, however, can operate in large numbers and be lost and replaced in equally large numbers. Putting fewer human beings physically in harm's way may be more effective militarily and produce better, ethical outcomes.

The future force should also be built around highly decentralized networks that move limited amounts of data rather than the highly centralized networks of today that must move tons of data. The reason current US military networks are so vulnerable to attack is because they are constituted around a small number of centralized hubs that opponents can easily strike. Having machines that can interpret the data that they collect themselves would enable the US military to move far less information around its networks and to distribute key network functions across large numbers of intelligent machines. Competitors would have a difficult time disrupting networks that lack large, vulnerable centers and that can physically reconfigure and heal themselves. These kinds of networks would likely stand a better chance of surviving on future battlefields, and they could enable human operators to remain in better communications with their intelligent machines.

Finally, the future force must be defined more by its software than its hardware. It must be, in every way, a digital force. This is a total inversion of how military power has forever been conceived. What traditionally wins wars is hardware. It is iron and steel. Hardware will still be important, but what will more likely win future wars is information. It will be the ability to build battle networks in which every military system can connect and collaborate with all others. And the capabilities most essential to success will be artificial intelligence, machine autonomy, cyber warfare, electronic warfare, and other software-defined technologies. These will enable human beings in combat to close the kill chain faster and more effectively than their rivals. In this way, future military hardware should be valued more as vessels for advanced software, similar to the way we view our mobile devices. What makes my iPhone special is all of the software inside that no one can see, whereas the hardware is just a low-cost platform that routinely gets expended and replaced.

I must emphasize that this future force will not be cheap. It cannot be purchased with a radically slashed defense budget. That is because the goal is not to replace one expensive system with one cheaper system but rather to replace it with a network of *many* cheaper systems. Over time, this should involve replacing legacy platforms with large networks of autonomous systems—a Military Internet of Things. It should also involve making trade-offs that cut across different military domains, perhaps replacing large warships with large quantities of land-based anti-ship missiles or replacing ground systems with networks of unmanned combat aircraft. The goal is to acquire whatever combination of smart systems adds up to a superior capability that enables humans to understand, decide, and act. As always, the focus must be on the kill chain.

If we are to pursue a strategy of defense without dominance, there is one additional insight that Americans must grasp: We cannot do this alone. Even changing the ends, ways, and means of the US military will be insufficient to create a favorable balance of power amid the continued emergence of a technologically advanced peer competitor in China. The United States needs capable allies and partners to succeed in the world, especially to balance Chinese power.

Trump is not wrong to demand that wealthy allies of the United States contribute more to our common defense. Indeed, friends who want America to help defend them if they are attacked have a special obligation to make themselves more defensible. And if they do not, they should not expect Americans to fight wars on their behalf that may have little prospect of victory.

Having higher expectations of our allies, however, should not be confused with deriding the value of having allies at all. This is one of the most fundamental problems with Trump's worldview, which he summed up in June 2019: "Almost all countries in this world take tremendous advantage of the United States," he said.[5] Trump sees America's allies solely as freeloaders and free-riders, and unfortunately, he has done more to damage America's relationships with our closest allies and to drive wedges between us than our adversaries ever could hope to achieve on their own.

What Trump gets wrong is that the United States does not have allies because we are suckers. We have allies because it benefits America. We want allies because it is better than being alone. We need allies because maintaining a favorable balance of power is not possible without them.

If we want our allies to be more capable and to share more of the burden of our collective defense, as we should, we should start by recognizing how the United States has intentionally contributed to our allies being less militarily capable. For example, we have often refused to sell offensive strike weapons and advanced defensive capabilities to frontline allies in Asia and Europe because we have believed that to do so would be destabilizing and provocative. We have also taken a limited view of the operational utility of those allies. As we have considered what we would want our allies, such as Japan, to do in the event of a regional conflict, US leaders have mostly relegated them to remaining in the rear, minding their own stores, and holding America's coat while our military moved forward to do the offensive heavy lifting on our own.

The United States has had good reasons for limiting the military power of our allies, not least a concern that more capable allies might be tempted to start or escalate fights that could implicate us in misguided conflicts. This is a legitimate risk. But the greater risk now is not that America's frontline allies are too militarily capable; it is that they are not capable enough and integrated sufficiently into meaningful roles in the US military's operational planning. Washington leaders pay lip service to the importance of allies. What we often convey through our actions, however, is that allies are nice to have, but if push really comes to shove, we prefer to do the hard things on our own. This must change for America to deny China military dominance.

That goal is simply unachievable without major operational and political support from our allies. We must require a lot more from both our allies and ourselves. Considering how fast a future war could start and escalate, America needs our allies to be capable of immediately defending themselves and us from any acts of aggression. We also need our allies to be willing to host significantly larger amounts of US military power than they do now, because America no longer has the luxury of commuting to future conflicts from stateside bases on multimonth deployment schedules. These expectations of our allies entail much larger political and diplomatic burdens, and our allies would bear those burdens not only for themselves but also for us—because the US military may not be able to defend effectively without their support and access to their territory.

The United States is headed into a future that will be as unsettling as it is unfamiliar, but we do not need to fear it. We can still manage to defend the people, places, and things we care about most. Even amid the erosion of our military dominance, America can avoid a future in which a peer competitor is able to consolidate its own position of military dominance. Achieving this more limited, defensive goal requires a wide-ranging reimagination of America's defense strategy, which is possible, but not optional. The main question is not whether the US military *should* change but whether we *can* change—and change fast enough.

ELEVEN

BUREAUCRACY DOES ITS THING

In 2019, Secretary of Defense James Mattis and Secretary of the Navy Richard Spencer tried to do the right thing. They wanted to free up room in the Navy's budget to invest in new capabilities, such as unmanned vessels like the Orca. And though they did not know how many aircraft carriers America would need in the future, they believed it would be fewer than today.

Mattis and Spencer set about trying to reduce the amount of money the Navy would ask Congress to spend on aircraft carriers in the 2020 fiscal year. One idea was to curtail the Navy's plan to purchase two new carriers at once, especially considering that the Navy was still wrestling with how it would defend these massive ships against a growing Chinese threat. This idea made strategic sense, but the Pentagon leaders eventually discarded it because canceling the "block buy" of two new carriers would have resulted in massive layoffs at the one American company that builds the ships. Once those workers were gone, there was no getting them back, which would harm the Navy's ability to build different kinds of ships in the future. The shipbuilders needed ships to build, so the carrier block buy was left untouched.

What Mattis and Spencer settled on instead was a plan to retire one existing carrier, the USS *Harry Truman,* halfway through its service life. The nuclear reactor that powers a carrier like the *Truman* can last fifty years, but it needs to be refueled after twenty-five years. By retiring the *Truman*, Pentagon leaders expected to save $3.5 billion on the refueling and $30 billion more that would otherwise go to operating and maintaining the *Truman* until the end of its life.

Things went downhill quickly. Congress was caught off guard. No one had explained in advance why the hard choice to retire the *Truman* was worthwhile or prepared Congress to make the decision. What's more, the new capabilities that Pentagon leaders wanted to invest in existed more on paper than in the water, which made it appear like they were trading a known quantity for unproven hopes. The Department of Defense was not prepared to make its case on the *Truman* about the relative capability and survivability of aircraft carriers, especially because it was proceeding with a plan to buy two new carriers at a total of more than $25 billion. Indeed, Navy sources soon made it known that they would be happy to keep the *Truman* in service.

The reaction was swift and brutal. Members of Congress from the states most affected by the *Truman* decision, backed by the companies, workers, and unions that would have benefited from that work, as well as the armies of lobbyists and consultants those interested groups retain to fight battles such as this, all mobilized to kill the department's plan to retire the carrier. Before they ever got the chance, however, just one month after the plan was released, Vice President Mike Pence visited the *Truman* in Virginia, where aircraft carriers are primarily built. In a speech on the deck of the ship, Pence proudly announced that the Trump White House, clearly with an eye toward the importance of Virginia in its upcoming reelection campaign, was reversing its own plan to retire the ship.

The audience went wild. Congress added the funding. And many of those future capabilities paid the bill. As a result, despite threats to aircraft carriers growing qualitatively and quantitatively worse, and though it is unclear how defensible US carriers are now, Washington will invest more than $30 billion to keep a twenty-five-year-old ship at sea until nearly 2050.

The case of the *Truman* is just a stark example of how difficult it is to make significant changes to America's military. It is also an example of something that happens all the time. Interested parties in the Department of Defense and Congress team up, with the strong backing of outside groups, to buy additional legacy ships, fighter jets, ground vehicles, and other military platforms. And more often than not, the money to pay for them is taken from equal, if not greater, priorities, such as the weapons and ammunition that those systems require, the ability to operate and maintain them, and most often, investments in future capabilities and technologies.

The battle lines for these fights are rarely drawn neatly between the executive and legislative branches, the two houses of Congress, or Republicans and Democrats. They often involve interest-based coalitions that transcend these institutional and partisan divisions. I used to facilitate these kinds of decisions all the time. Sometimes it was the wrong thing to do, but members of Congress or powerful groups in the Department of Defense wanted to do it anyway. At other times, however, the Department of Defense made the wrong decision in its budget request and Congress was right to overturn it.

At the center of it all is what is known in Washington as the budget process, the annual practice of how the Department of Defense and Congress, with the active involvement of all manner of interested parties—not just defense industry but also veterans' organizations, unions, environmental groups, state and local governments, and many other stakeholders—determine how America will allocate its annual defense spending, which totaled nearly $700 billion for fiscal year 2020. Think of the budget process as the opposite of the kill chain. Whereas the kill chain is supposed to be fast, meticulous, and uncompromising in its precision, the budget process is slow, tedious, unruly, and defined by messy, imperfect compromises. And yet both are essential.

The budget process matters because, as the old saying goes, budgets are policy. Even $700 billion cannot buy everything that everyone wants. Leaders have to make choices, and those choices require trade-offs. Buying this means not buying that. Investing in a new capability for the future often means giving up something today to pay for it. There is no way around it. If you want to know what leaders in Washington really value, what they say matters a lot less than what they spend money on. Spending is what reveals their true priorities—what matters most.

Do not think that these Americans are stupid, malevolent, or venal. Most are doing their best to do right as they understand it. They are elected officials trying to stand up for their constituents, who build or operate amazing capabilities for the US military and want to keep their jobs. They are military officers who believe in the utility of their current systems and that having more of them will make their troops more likely to succeed in their missions and return home safe and sound. They are business leaders who have a responsibility to do right by their workers and to make money for their shareholders, most of whom are ordinary Americans.

The problem is not the people involved. Nor is it the mere existence of what John McCain used to call "the military-industrial-congressional complex" or what Donald Trump and others have more recently called "the swamp." After all, that system is not going anywhere. The bigger problem is how power and incentives have come to be structured over many decades within America's defense establishment. That structure overwhelmingly favors the present at the expense of the future. Most people are rewarded and punished in the budget process based on how well they look after the needs of today, not the needs of tomorrow. That is largely what wins congressional votes, military promotions, and corporate bonuses. These incentives generate a powerful reluctance to do things differently, take more risk, and move more urgently.

That is why America got ambushed by the future so badly a few years ago. And most of the conditions that contributed to that systemic failure unfortunately still exist—a defense acquisition system that has been optimized for risk aversion and cost accounting, not rapid technology development at scale; a defense industry that has become increasingly consolidated and closed to new entrants; a breakdown in the relationship between the national security and technology communities; and a broader failure of imagination about America's rapidly diminishing military dominance. Our nation's defense bureaucracy requires broad consensus to do nearly everything, especially to enact changes as sweeping as those now required, and consensus is perhaps the single hardest thing to come by in Washington today—or, for that matter, in America itself. If this transition fails—and the odds of that are unsettlingly high—there will be many reasons for that failure, and most of them will revolve around the budget process.

The budget process is actually bookended by two other bureaucratic processes that are also essential to how America builds and buys its military. What precedes the budget process is "the requirements process," and what follows it is "the acquisition process." If all of this talk about process sounds exceedingly bureaucratic, it is.

The requirements process is how the Department of Defense determines what constitutes a "validated" military capability to develop or buy. The reason for this process is sound—to ensure that the most pressing needs of joint warfighters, not parochial preferences of individual components, get funded and procured. The problem is that the process that validates requirements can drag on for months, even years, leaving troops in need without cutting-edge or even effective capabilities. The requirements process can also become dominated by consensus, generating tons of questionable and frequently changing priorities to which no one is willing to say no. This has frequently resulted in requirements for "unobtainium," which eventually collapse beneath their many contradictions.

A bigger problem is that the requirements process can be a black box where defense bureaucrats try to micromanage the exact parameters of the weapons they know about while unintentionally shutting out the many new technologies they do not know about. The process can become divorced from a realistic outlook on what is operationally necessary at present and what is technologically possible in the future. At worst, it prevents military operators from even being able to experiment with new technologies, let alone field them. I used to joke when I worked in the Senate that if a commercial company developed giant robots that shot lasers from their eyes, the Pentagon would reject these super weapons and claim that there is no validated requirement for such things.

What comes after the budget process is the acquisition process, which occurs after the Pentagon has established its requirements and decided to budget money for them. Entire books can be written (and have) about our defense acquisition system. Criticisms of how America buys weapons and other military things, as well as attempts to fix that system, began soon after the Continental Army took the field in 1776, and they have continued ever since—for good reason.

A good example of how defense acquisition can go wrong is the Army's attempt to buy a new pistol a few years ago. It issued a request for proposals that ran over 350 pages of cumbersome details and envisioned years of costly development and testing before soldiers would ever get a new sidearm. Even Army leaders were surprised. They learned about it when McCain and I told them, and then they were as outraged as we were. "We're not figuring out the next lunar landing," said an outraged General Mark Milley at the time, when he was chief of staff of the Army. "This is a pistol. Two years to test? At $17 million?" he vented. "You give me $17 million on a credit card, and I'll call Cabela's tonight, and I'll outfit every soldier, sailor, airman, and Marine with a pistol for $17 million. And I'll get a discount on a bulk buy."[1]

Weapons, of course, need to be developed and tested deliberately. The problem is that this process has become so bureaucratic, so risk averse, so filled with people who can say no, so inclined to develop, test, and buy different things in the same ways, that too often we just make simple things hard, like buying a new handgun. Pentagon leaders have every authority they need to get our military the best possible technologies more rapidly, especially after the hundreds of pages of defense acquisition reform legislation that I helped to write for McCain and pass into law.

The issue is not a lack of authority to go faster or take more risk, but that those who must exercise those authorities, bear those risks, and be accountable for the outcomes rarely use the authorities they have. The handgun debacle was a classic case of decision making by committee. After every part of the bureaucracy weighed in with what were possibly reasonable concerns, the end result was unworkable. Milley could have decided to go faster or buy the weapon differently, which the bureaucracy would have found risky, but he could have used his authority as a senior leader to do so because he was ultimately accountable for getting soldiers the weapons they need when they need them. Milley never made that decision because he never knew it was being taken. This is the deeper problem with the acquisition process. This happens all the time. Those who have the authority to do things differently rarely use it, and those who do make decisions often lack the authority and incentives to make riskier decisions to get better outcomes. This happens all the time.

The requirements process and the acquisition process often impede good outcomes. But the bigger problem is that we spend too much money on the wrong things and not enough on the right things. And that has more to do with the budget process.

How we spend money begins with what the Department of Defense calls "programming," which is how the world's largest government bureaucracy decides exactly what it wants to request money for from Congress. The Pentagon builds its budgets in five-year plans, much as the Soviet Union once did. This is called the defense "program," and much of its cost is carried over from one year to the next. Most people stay in the military for more than a decade. Many procurement programs take just as long or longer. Once the Pentagon starts paying for people, places, and things, it has to keep paying for them. This means that the majority of the money that the Department of Defense plans to get in future years has already been obligated by past decisions. And once those programs get started, it is incredibly difficult to stop them, because of how many stakeholders in and out of our government benefit from continuing them at all costs.

Of the limited future money that remains unspoken for, the process to plan how to spend it begins nearly two years before the Pentagon actually receives a dollar of that money from Congress. This means, for example, that Pentagon planners have to start predicting in January of the current year how they should spend money in the fiscal year that begins in October of the following year. In that gap of time, entirely new technologies are developed. Brand new companies are founded. And the Pentagon cannot plan to take advantage of any of them, so it programs its future money toward capabilities and technologies that it knows about now, which makes it exceedingly difficult to be dynamic, adaptive, and responsive to unforeseen conditions.

This process already constrains the Pentagon's ability to invest in future capabilities, and Congress rarely makes it any easier. For years, Congress has consistently limited the military's flexibility to spend money, often not without reason. But the result is that Congress directs how much and in what ways the Pentagon can spend money on each of its "programs of record," such as the M1 Abrams main battle tank, as well as on the multitude of weapons, spare parts, and other subsystems that comprise each of those programs. What's more, if the Pentagon wants to shift, or "reprogram," any of this funding for other purposes, it often requires permission from four different congressional committees, and the total amount of money that the Pentagon is allowed to reprogram in a given year is less than .009 percent of its budget. The idea of giving the military accounts of money that it can spend more flexibly on emerging technologies is especially unpopular among congressional appropriators, who view them as "slush funds."

All of this makes it difficult to get new technologies to the military quickly, and that is further exacerbated by the structure of power in the Department of Defense. It is often assumed that the Pentagon is the ultimate hierarchical system, where leaders at the top are all-powerful and can direct everyone to do everything they wish. In reality, however, the structure of power reflects how the Department of Defense has been built over many decades—from the bottom up.

Before there was one secretary of defense, power resided with the heads of the Army and the Navy. Prior to that, power largely resided with the leaders of the different branches of those services, such as infantry or armor (in the case of the Army); surface warfare officers or aviators, to say nothing of the Marine Corps (in the case of the Navy); and fighter or bomber pilots (in what became the Air Force). As each layer of bureaucracy was added to govern the sprawling defense enterprise, some power shifted to the top. But much of this was power on paper. In reality, most power still remains at lower levels, concentrated ironically in what are known within the largest non-democratic institution in America as "communities of interest."

The Pentagon develops its budgets, for example, from the bottom up. Those core military institutions that were once all-powerful—and, in subtle ways, still are—set in motion thousands of decisions in the budget process totaling hundreds of billions of dollars before the Pentagon's senior leaders ever see them. And many of those choices get rubber-stamped each year because those senior leaders simply cannot process all of them. What this means, in practice, is that countless decisions affecting enormous amounts of defense spending are made by entrenched parochial interests spread around the Department of Defense that have neither the authority nor the incentive to make bold moves that change America's defense program. This leads the Pentagon's many communities of interest to view their senior leaders, who come and go every few years, as tantamount to part-time employees who are not around long enough to really matter—or, as a friend in one of those communities once put it to me, "the Christmas help."

Most of the incentives that govern the Pentagon's bureaucracy favor the past over the future. Military servicemembers are only in a given job for a few years before they rotate to another one. In that short time, they are rarely rewarded for rocking the boat or raising problems up the chain, least of all when their complaints regard the failure of their own institutions to do new things or adopt new technologies for which few people as yet see a need. Such disruptions are more often viewed by the powers that be, who manage military careers, as a reason to doubt whether a person is a team player who deserves a top job in the next promotion cycle. Those who are rewarded are people who shepherd the existing programs of their respective communities of interest through the budget process with as little change as possible. If future leaders do not get what they need, they cannot complain about it today. But current leaders sure can, and do—loudly.

The very structure of the Department of Defense can even be a disincentive to right thinking. The most important thing that the US military must do is close the kill chain, but the Department of Defense is organized around military services and the platforms that they often allow to define their identities. The Navy fixates on "ship count." The Air Force fixates on its number of squadrons. The Army fixates on its "end strength," the number of soldiers in its ranks. And the Marine Corps has traditionally fixated on amphibious ships. Individual members of each service are the first to admit that this kind of thinking is backward, but the institutions still engage in it because the incentives to do so are powerful: counting people and things, especially traditional platforms, is an effective way to compete for money in the budget process.

The result, however, is all manner of misplaced priorities and misguided decisions for the US military overall. The Army, for example, is set up to think about closing the kill chain primarily on land with its own ground forces, whereas the Navy is set up to think about doing so primarily at sea with its own maritime forces. But if the fastest and most effective way to close the kill chain against enemy ships is with land-based capabilities—or, better yet, a mixture of different services' capabilities—those kinds of solutions do not arise easily or naturally in the Pentagon because the institution is structured more to resist them than to produce them.

None of this makes it easier for US military institutions to exploit the full potential of new technologies. Instead, the incentives that dominate those institutions often lead them to see the value of new technologies primarily in how they can improve traditional military systems, not create new kinds of capabilities and new ways to operate them. For example, despite decades of progress in unmanned aviation, both the Navy and Air Force are planning to spend billions of dollars to develop new, manned fighter jets that they expect to deliver to the force many years from now. Both services are also developing autonomous aircraft, but they are limiting them to missions centered around traditional, manned fighter jets—refueling them, in the case of the Navy, and defending them, in the case of the Air Force. Those are important roles, but they are hardly the most compelling roles that autonomous aircraft are capable of performing now, let alone in the years to come.

There are only a handful of leaders inside the Pentagon with the power to bend the budget process decisively in favor of the future, most importantly, the secretary and deputy secretary of defense, but they too are overwhelmingly consumed by the present. The secretary of defense sees and hears from literally dozens of people every day, and nearly every single one of them is calling about the present, not the future. When the White House calls, it is usually related to some ongoing foreign policy crisis or military operation. When foreign leaders call, it is often for the same reasons. When members of Congress call, it is invariably because they want urgent attention to some current problem involving their constituents. And when other Pentagon leaders come calling, it is rarely to contemplate emerging technologies and the future of warfare. These kinds of present challenges have a major bearing on the day-to-day effectiveness of the Pentagon's senior leaders, but they are all-consuming and leave little time to think about and plan for the future.

All of the incentives that push the Department of Defense in favor of the status quo are just as strong, if not stronger, in Congress. Members of Congress cannot ignore the things and issues that are likely to win or lose them votes. Nor can they be faulted for this. It is the reality of a democratic system. What makes this problematic for the future of the US military is that the future does not vote, but the present certainly does. And throughout the budget process, members are swamped by constituents who have problems now.

A clear indication of what members of Congress really care about, in addition to how they spend money, is how they spend time. They cannot get back either once they spend it. For four years, I had one of the few jobs in Congress that afforded a clear view into what America's elected representatives really cared about in the defense budget process. I never once had reason to doubt whether any member of Congress wanted what was best for America and its military. And many were capable of distinguishing between parochial issues and national interests.

However, what got most members of Congress most riled up in the budget process was not the need to spend more money on new technologies or to push the military to change for the sake of the future. It was almost always the present issues closer to home: a problem facing their local National Guard unit or an environmental or military construction priority for their constituents. Or it was a desire to spend more money on a current program that was built or based in their state or district, which inevitably meant taking money from something else.

Indeed, one of the biggest, most time-consuming battles I was pulled into during my years in the Senate was the push by a committed group of members of Congress to prohibit the greater sage grouse, lesser prairie chicken, and burying beetle from being listed as endangered species. I will spare you the reasons why they wanted to do this, which were actually rooted in legitimate concerns. But suffice it to say that trying to resolve what we called the "critters" issue each year occupied countless hours of McCain's and other congressional defense leaders' time—time that could not be spent on the future of the US military. And it was not just the critters. That was just one of the many urgent but lesser priorities that consumed us each year.

It is often assumed that the actions of members of Congress are directly linked to intense lobbying by special interest groups, especially defense companies. This is too simplistic a view. Defense lobbyists can play a significant role in mobilizing people in favor of or against something, it is true. And as the few big defense companies have become more fixated on their quarterly earnings, they have become even more focused on trying to get as much as possible for themselves in the budget process each year. But it is not as if members of Congress need lobbyists to tell them that they have important constituents who want to keep building legacy military systems. I rarely saw members of Congress spend considerable amounts of time fighting for or against something just because lobbyists requested it. Members would do so for their own self-interested reasons.

Defense lobbyists are a convenient scapegoat. But the real problem is not that a handful of big defense contractors have a loud voice in the budget process. The real problem is that so few defense companies are left in America after decades of defense industry consolidation, that so few of the remaining companies are leaders in emerging technologies, and that those which are doing this futuristic work for the US military have little to no voice in the budget process. None of this is the fault of defense lobbyists. But all of this makes it much more likely that the future will lose in the end.

The truth is that Congress has considerable power to correct the failings and oversights of the Department of Defense and the defense industrial base, but Congress too often uses its awesome powers for things that just do not matter that much to the future effectiveness of our military. It is hard not to think that this is related, in some way, to the significant reduction in the number of members of Congress with military experience, which is roughly half of what it was thirty years ago, which contributes to a growing unfamiliarity with the US military among the very people charged with overseeing it. This distancing can contribute to two opposite but equally deleterious behaviors: a hostility toward, and desire to purify, the less egalitarian aspects of military culture, on the one hand—and on the other, a kind of unthinking deference to those in uniform that can often lead civilians who never served in the military to think that they should not, and indeed cannot, question the core business of America's professional military class.

In addition to lacking military experience, most members of Congress and many congressional staff also lack technical knowledge and experience in technology. The same goes for most civilian and military leaders in the Department of Defense, where technical backgrounds (as opposed to warfighting prowess) are seldom grounds for promotion to the highest ranks of military service. In Congress, less than 1 percent of members have studied computer science, and few have meaningful experience working in the technology industry. Of course, the background experience of members of Congress is not a prerequisite for America to prioritize investment in new technologies for its military, but more leaders who understand these technically complex subjects can only help.

It is easy for America's political and military leaders to become enamored of emerging technologies now, when these technologies are largely unthreatening to the traditional systems and ways of doing business. But in time, emerging technologies will no longer simply enhance traditional military systems. They will threaten to replace them altogether. Indeed, it has been estimated that artificial intelligence could automate 45 percent of the tasks in the US economy.[2] That figure might be even higher for the US military, which is a decade or more behind the commercial world in its adoption of many modern information technologies. The shift to emerging technologies in place of existing systems could happen much faster than most people in the defense establishment realize, and the backlash could be severe. After all, fighter pilots are no more eager to lose their jobs to machines than factory workers are.

The current revolution in military affairs will inherently lead to a broader economic, social, and political disruption—which is profoundly challenging for any democratic society to manage, no matter how militarily necessary it is. This disruption will impact the fortunes of major companies that build legacy military equipment. It will call into question the work now being done by hundreds of thousands of Americans in uniform. And it will threaten the livelihoods of potentially millions of Americans who have long derived a sense of pride and dignity from the exquisite work they perform on behalf of the nation's defense. As the numbers of those disaffected and displaced by emerging technologies grow, they will find ample opportunity in America's defense budget process to resist the pace of disruption. We should sympathize with their reasons for doing so, even as we recognize the challenge it poses.

Seeing through change of this magnitude—from a military standpoint, let alone a social, economic, and political one—would be exceedingly difficult in normal times. And these are hardly normal times. We need leadership in Washington that can manage the political risks and fallouts associated with nothing less than a generational change of America's national defense. What Washington is providing instead is a level of political dysfunction that is unique in modern American history and that is bleeding into national defense in a way that makes everything harder. This problem is bigger than Donald Trump, but he is at the center of it.

This is not to say that the Trump presidency has been uniformly bad for our military. It has not. But in a broader sense, the past four years have been tumultuous ones for national defense. Trump has routinely dragged the US military into his broader political agenda, sending troops to the southern border largely for symbolic reasons and raiding billions of dollars of military funding to pay for the border wall. After prioritizing strategic competition with China, Trump then proceeded to heighten tensions with Iran and Venezuela in ways that diverted military funding and focus. His personal attacks on Amazon CEO Jeff Bezos have hindered the Pentagon's attempt to procure enterprise cloud computing services, which are a prerequisite for any serious development of artificial intelligence. These and other present burdens all come at the expense of our military's future.

At the same time, many senior civilian jobs in the Department of Defense have gone unfilled for long stretches of time; many seats remain empty. In this absence of leadership, many hard choices necessary to move the US military into the future simply are not made, because career civil servants and military officers do not have the authority to make them. The result is that much of the Pentagon has been standing still, which really means falling behind.

Unfortunately, the Trump presidency has only exacerbated the state of dysfunction and distraction that was already consuming Congress. In the nearly one decade that I worked in the Senate, I watched a radicalization of political discourse occur that was frightening in its speed and severity. Republicans shifted sharply to the right, and now Democrats are reacting by shifting sharply to the left. There are many reasons for this, but the result is a hollowing out of the political center and a total incapacity to come together to do much at all. And increasingly, some of the best and brightest members of Congress are just opting out altogether and leaving.

Even basic tasks that used to be routine bodily functions in Congress, such as passing a federal budget, have become nearly impossible. Indeed, over the past ten years, Congress has managed only *once* to pass spending legislation for the Department of Defense by the start of the fiscal year. When Congress fails to do its job in this way, it passes a "continuing resolution," which requires the military to spend money on only the things it spent money on the prior year. Not only does this waste billions of dollars in misallocated resources, it literally locks the military into the past and prevents it from implementing its plans for the future. This is how the Department of Defense has spent nearly one thousand days of the past decade.

The US military now plans to start each fiscal year *without* an appropriation of funding. Pentagon planners painstakingly negotiate contracts and structure programs to avoid critical payments in the first quarter of each fiscal year so they do not end up in breach of contract when they inevitably get stuck on a continuing resolution. Even then, problems arise. When Congress failed to pass a budget for six months at the start of the 2018 fiscal year, for example, the Navy had to renegotiate roughly ten thousand contracts, which senior Navy leaders estimated cost them roughly $5.8 billion in wasted buying power. That could have bought three destroyers.

And yet this is exactly how the Department of Defense began the fiscal year for 2020—once again on a continuing resolution with no clarity of when it would end. The reason was the fight between the

White House and Congress over funding for a southern border wall. But if it had not been this fight, it would have been something else, just as it has been each year for the past decade. This is how previously deviant behavior has now become all but normalized. It cannot be blamed narrowly on the behavior of Trump or Congress, Democrats or Republicans. These are just symptoms of a deeper failure of imagination and leadership in Washington—a failure to grasp the true stakes for America in the strategic competition with China and to subordinate typical political gamesmanship to the broader need to prepare our nation to succeed in the future.

This has not happened for lack of clear warnings from senior US military leaders about the consequences to America's men and women in uniform. On April 10, 2019, Admiral Scott Conn, the Navy's director of air warfare, testified to McCain's old committee in the Senate, shortly after I had left. "I had a meeting with the Top Gun commanding officer and two lieutenants who are on his staff," Conn said. "We went over in a classified setting the pacing threat. We went over what we had planned in 2018, what was budgeted in 2019, what we are requesting in 2020, where we are going in 2021. If we go back to a continuing resolution, that stuff gets blown up," he continued. "And what we are transmitting to those lieutenants is that we are not committed to winning."[3]

Listening to Conn's plea, I was reminded of something I often wondered during my time in the Senate. Our political leaders, in both the executive and the legislative branches, are being told in no uncertain terms the damage they are doing to America's military and its future ability to defend the nation. Is the problem that they are not paying attention? Or do they not care?

TWELVE

HOW THE FUTURE CAN WIN

In February 2018, Secretary of Defense James Mattis and Secretary of the Air Force Heather Wilson tried to do the right thing. They proposed to Congress, in the Department of Defense's budget request for the upcoming fiscal year, to abandon a prior Air Force plan to buy a new version of a thirty-year-old aircraft called the Joint Surveillance and Target Attack Radar System, or JSTARS. This system had been one of the central capabilities developed as part of Assault Breaker in the 1980s. The plane itself is a variant of a standard commercial aircraft, but it hosts a powerful radar that can find moving targets on the ground, such as enemy tanks and vehicles, and then pass that targeting information to other systems that can close the kill chain.

JSTARS had played a critical combat role in all of the wars of the prior decades, from the Gulf War in 1991 to the Balkans campaign to the invasion of Iraq in 2003, and the Air Force had been planning for years to recapitalize the entire fleet for billions of dollars. By 2018, however, the world was looking different. Air Force leaders had belatedly but nonetheless realized that JSTARS would not survive in a conflict against China or Russia. It was a business jet. It was not stealthy and had no means of defending itself. Even a new version of JSTARS would be found by Chinese or Russian radars and then shot out of the sky by their advanced missiles.

The new plan called for disaggregating the JSTARS mission. Rather than putting all of its eggs in one vulnerable platform, the Air Force wanted to develop a network of unmanned aircraft and satellites that could find moving targets and then fuse all of their streams of intelligence into a common picture of the battlefield. This new capability would be more resilient and survivable than JSTARS. Instead of one defenseless aircraft, adversaries would have to find and attack multiple systems spread across wide swaths of air and space. This kind of Military Internet of Things was the right idea, and I became a strong supporter of the new plan on Capitol Hill.

There was a big problem with all of this, however. Powerful members of Congress were counting on the new JSTARS aircraft to be based in their states. Influential companies stood to make billions of dollars over decades building and maintaining that new JSTARS system. And these forces mobilized quickly to try to kill the Air Force's revision of the JSTARS plan in the budget process.

The reason the Air Force was ultimately successful in shifting to a new program was that it did a number of key things right—things that the Navy, for instance, failed to do with its plan to retire the *Truman*. The Air Force devised a political strategy. It engaged candidly and early in the budget process with key stakeholders in Congress. It provided detailed information about the threat to JSTARS and how the new program would be better. It made sure the defense industry was informed so the companies that thought they might benefit under the new plan could conduct their own private lobbying on Capitol Hill. The Air Force also threw a bone to those members of Congress who stood to lose the most by committing to base elements of the future program in the same state that would have hosted JSTARS, thereby turning potential spoilers into strong supporters of the new plan.

Even then, this was one of the single most contentious fights in the budget process that year. The Air Force's plan faced strong and vocal opponents both in Congress and outside of it. Indeed, of the more than eight thousand individual provisions that the House of Representatives and the Senate had to resolve in their annual defense legislation that year, what to do with JSTARS was literally the last decision to get made, and it ultimately came down to the four bipartisan leaders of the defense committees. I was in that room. If not for the tenacity of some of those leaders, the decision could easily have gone the other way, and Congress could have required the Air Force to spend billions of dollars on an outdated capability that would not survive in a future war.

One way to look at the JSTARS saga is as representative of everything that is frustrating and even rotten with the "military-industrial-congressional complex" or "the swamp." A good idea rarely wins on its merits alone. Its success too often comes down, instead, to the trading of favors and the political dark arts.

I take a different view on my JSTARS experience, and it is this: If the future is going to win, it will have to win *inside* our current system. It will have to win in a system comprising parochial military services, self-interested companies, and largely distracted political leaders—all of whom will continue to be consumed more by present concerns than future ones. Those are the people, interests, and political realities that matter, and none of them can be hand-waved away. This system makes it exceedingly difficult for America's military to change, or change quickly, but as I have said from the start, I would not be writing this book if I thought all hope was lost.

There is good news and bad news. The bad news is that those who want to change America's military face huge obstacles and opposition. But here is the good news: The United States is not lacking for any of the key elements that a change of this magnitude requires. We have plenty of money. We have amazing, world-leading technology. We have creative and talented people. If America lacked any of these elements—which many of our foreign competitors do—the prospect of us adapting for the future would be much bleaker.

America's main problem lies in Washington. It rests in the choices and decisions we have made. It encompasses our increasingly dysfunctional political system that too often appears unwilling and unable to perform even the basic functions it is supposed to do, let alone the more ambitious and vital tasks that it needs to do now. And it relates to our failure to imagine how our national defense can and should change and our lack of urgency to bring this about. None of these issues is beyond our control to make better. Ultimately, it comes down to two things: incentives and imagination.

America's defense problems result from incentives that Washington has created over many decades—incentives that have favored better legacy platforms over integrated networks of faster kill chains, familiar ways of fighting over new ways of war, offense over defense, present needs over future ones, hardware over software, acquisition compliance and cost accounting over rapid development of new capabilities, traditional defense companies over new technology developers, industrial consolidation over a diversified ecosystem of defense technology. These outcomes are the direct result of how America's defense establishment has defined its priorities, spent its money and time, and rewarded and punished its people. We have gotten what we have paid for. And if we want different outcomes, we must create different incentives.

This will not happen quickly or easily. It amounts to a sweeping restructuring of our defense establishment. But it is entirely within our power to do. The question is how.

To prepare America's military for the future, we must *want* to change, and there is reason to be encouraged. The impetus to change is stronger now than it has been for a long time among many defense leaders, especially the US military itself. More "military mavericks" are rising to the fore, and they want to use their unique authority and legitimacy to change their institutions. The JSTARS decision, for example, was driven by Air Force officers and was embraced by their civilian leaders. More recently, the Army's two highest-ranking generals, together with its top two civilian leaders, convened what became known as "night court," regular after-hours meetings in which the four of them sorted through every early-stage development program in the Army to identify—and cut—the ones that did not align with the new defense strategy. The result was $25 billion in savings over five years that the Army asked Congress to shift toward higher-priority programs.[1]

Not to be outdone, in July 2019, just one month after being confirmed as Commandant of the Marine Corps, General David Berger laid out a vision to change his service that was bolder than anything put forward in a generation.[2] For decades, it had been holy writ for the Marine Corps that it required thirty-eight amphibious ships, which are better versions of the platforms that sent Marines ashore at the start of the Korean War in 1950. Berger jettisoned that requirement, suggesting that these big multi-billion-dollar ships might not survive against technologically advanced militaries. He sought instead to redesign the Marine Corps, less to project power than to counter the ability of competitors to do so, and he directed that the future force be built around smaller, lower-cost, more expendable, and more autonomous systems.

Developments such as these are encouraging, but they are only initial steps in what will be a long and arduous journey. Washington is more than capable of squandering the opportunity it now has. We have done it before. To translate this current opening into lasting change, America's leaders must create a new set of incentives that can enable the future to win within a political system and a budget process that will remain dominated by the needs of the present.

Carrying through a change of this magnitude is something that senior leaders must own. They are the only ones who can do it and should do it. Plenty of officials and staff at lower levels want to initiate sweeping changes to US defense policy, and they often know better than their leaders what is needed. But Pentagon bureaucrats and congressional staff have neither the authority nor legitimacy to initiate the kinds of changes that are required, which will disrupt the livelihoods of working Americans and the functioning of the US military. That power and responsibility rest only with those leaders who have been elected by the American people or confirmed by the Senate for positions of public trust within our government.

Restructuring the incentives of America's defense establishment is not something that senior leaders in either the Department of Defense or Congress can do on their own. They must do it together. These leaders are equal under the Constitution and must deal with each other as such. They will inevitably check and balance each other's power, as they should, but what America needs now is less adversarial checking and balancing and more making of common cause. This means that Pentagon leaders must resist the temptation to treat their counterparts on Capitol Hill like children—listening and nodding while they talk but then refusing to include them as grown-ups in meaningful deliberations about the future. This also means that congressional leaders must resist the temptation to act like children—grandstanding when it suits while shirking hard decisions that involve real political risk and pain—and refrain from micromanaging the military simply because they can.

For the future to win, senior leaders in the Pentagon and Congress must forge a relationship of closeness and transparency to a degree that could make each deeply uncomfortable. But in the absence of this kind of partnership, which is most of the time, a wide rift opens between America's civilian defense leaders that makes it easier for other self-interested actors, from the military services and defense industry to special interest groups and foreign partners, to play each side off the other. By closing that gap between them, senior civilian leaders in the Department of Defense and on Capitol Hill can present a unified front to all other stakeholders and create a shared set of expectations about what is required for the future.

This partnership is also necessary for reasons of time. Senior leaders in the Department of Defense stay in their jobs for only a few years at most—sometimes less than that in recent years. This is why the Pentagon's communities of interest often view them as the Christmas help. Members of Congress, however, may remain in their jobs for decades. By the time John McCain passed away, he had been in the Senate for thirty-one years. There are drawbacks to this kind of political longevity, to be sure, but there are also huge benefits. While civilian and military leaders in the Department of Defense come and go, members of Congress can provide the institutional continuity and impetus to sustain big, multiyear changes to America's military.

This is exactly what is needed now. Getting from the military we have to the military we need will be a long transition. It cannot occur all at once. There are no technological miracles or *deus ex machina* to save us. Betting on such immediate deliverance is precisely how we got stuck with so many "transformational" procurement debacles over the past two decades. The only way to succeed is how Admiral William Moffett got the Navy to embrace aircraft carriers between the world wars and how General Bernard Schriever developed intercontinental ballistic missiles during the early Cold War: rapid but incremental pursuits of transformational goals.

I am confident that the next Moffett and the next Schriever are serving in the US military today, and senior leaders need to create incentives that empower them. This starts with defining the problems that need to be solved in more exacting detail. Schriever succeeded because he knew exactly what problem his leaders wanted him to solve: deliver a nuclear weapon on a missile to the other side of the planet in minutes. Defense reforms tend to fail when they cease to be anything more than vague buzzwords. At one point those buzzwords were *transformation* and *revolution in military affairs*. Now they are concepts such as *great-power competition, defense innovation, multidomain operations,* and *joint all-domain command and control*. Military mavericks cannot do much with these vague ideas. If senior leaders do not define their top problems more clearly, defense buzzwords actually become obstacles to real change because the bureaucracy simply rebrands everything it has long been doing using these new terms.

To avoid falling into this trap, senior leaders in the Pentagon and Congress need to get on the same page about how to translate generalities such as *great-power competition* into specific operational problems that military mavericks and engineers can understand and solve. Further, the problems should not be framed around the prerogatives of specific military services, such as the Army or the Navy, or around the roles of military platforms, such as fighter jets or combat vehicles, or around military domains, such as air, land, or sea. They should be framed around the kill chain—how to gain better understanding, make better decisions, and take better actions faster and more often than specific military competitors.

Denying China military dominance is a clearer definition of the US military's most pressing problem, but we have to be even more clear. Former deputy secretary of defense Robert Work has suggested that defending the people, places, and things that matter most to America in the Asia-Pacific region requires the ability to deny China's means of projecting power and committing acts of aggression. More specifically, Work assesses that US and allied forces would have to close the kill chain against 350 Chinese ships during the first three days of a conflict.[3] To do so means understanding where all of those ships are as they move across a vast expanse of ocean, sharing that information in real time with decision makers, and taking thousands of actions in an environment where it will be very hard to hide, communicate, and project power very far.

This is an extremely difficult operational problem to solve, and there is a host of others just like it. Many of those problems will relate to China, but many will relate to other challenges that the United States must build forces to address. Translating general challenges and objectives into clear operational problems is necessary across all of the threats that the US military must defend against, because it enables military mavericks and technologists to devise practical solutions. We must then demonstrate some of those solutions for our competitors, because that is actually how we deter war. By sowing doubt in the minds of our opponents about whether they could gain the benefits of war and aggression, we can make them less inclined to try.

Translating vague, buzzwordy goals into clear operational problems is necessary for US leaders to create the incentives that can generate better, more relevant capabilities for the US military much faster. When I worked in the Senate, private industry regularly complained that the Department of Defense did not know what it wanted. It asks for one thing today and the opposite tomorrow. After being burned enough times by the military's uncertainty, major defense companies are reluctant to invest much of their own money to develop new technologies and prefer to develop only what the government says it wants and is willing to pay them to build.

This is not an unreasonable stance. But the reality is that people who build militaries face the same problem that Yogi Berra summed up so well when he said, "It's tough to make predictions, especially about the future." If I had to ask for every technology that I believe I need, I would not have an iPhone today—or, for that matter, most of the other commercial products that I cannot live without. It is not possible for the US military to know exactly what it needs for the future. But it is possible to create better incentives that stimulate the development of new capabilities and produce novel solutions to our most pressing military problems.

Establishing more effective incentives will not happen through the traditional requirements process or through a bottom-up budget process that allocates money to lower priorities first and forces the top priorities to fight over the remaining dollars at the end. Instead, senior leaders need to create incentives that get the latest technologies into the hands of military operators quickly and allow them to experiment—let them learn through trial and error what works, what capabilities they could have and would want, how they might use new technologies to operate in new ways, and which of those new capabilities and concepts can close the kill chain most effectively. The best way to do this is to let key stakeholders compete against each other.

The Department of Defense rarely uses mission-focused competitions to identify the best solutions in the way that has proven so effective in the commercial world. Indeed, that is how machine learning has improved so dramatically in the field of computer vision in recent years: researchers and developers have come together over and over again to see whose algorithms could more precisely and routinely identify people and objects in images, and they all competed tooth and nail to be the best. The same thing can and should happen in national defense.

Senior leaders in the Pentagon and Congress should set aside a large sum of money every year at the start of the budget process and then hold competitions to determine who has the best solutions to the US military's highest-priority operational problems. These competitions must be real-world events featuring operationally relevant problems, like Work's scenario involving the real-time targeting of 350 ships, albeit on a smaller scale. These competitions should be open to the military services, defense industry, technology companies, government laboratories, and anyone else who can bring real solutions to the table—not PowerPoint presentations but working components of an integrated battle network. The goal should be to evaluate which capabilities generate superior understanding, decisions, and actions under highly dynamic conditions, and do it all faster and more frequently than the US military can today. And here is the most important part: we really have to reward the winners by funding and fielding their capabilities at scale.

It should not matter who wins these military competitions. The best solutions may all come from one military service. They may all come from a household-name company or a start-up that no one has ever heard of. The best solutions might rely almost entirely on the cyber domain and require few traditional platforms at all, or traditional platforms may win hands down. The winner might not even be a new capability at all but rather a new concept for using existing capabilities more effectively, such as *blitzkrieg*. It should not matter.

The goal should be to determine what closes the most kill chains fastest, not which is the best military service, platform, domain, or other thing. And to the victors go the spoils.

The amount of money that could be made available to fund and field the most effective military capabilities might realistically be only a fraction of the defense budget. But even 5 percent of that budget is $30 billion. That is a strong incentive that would attract the top performers and make it a priority for them to compete and win.

The goal should be to take advantage of the many deep-seated rivalries in the defense establishment. These rivalries are inherent between military branches and between companies, and they can be problematic when they lead to parochial solutions trumping joint or national solutions. But in their desire to mitigate these rivalries, senior defense leaders too often defuse them altogether, making the problem worse. For example, to try to prevent the military services from fighting with one another, leaders give each service an equal portion of the budget, regardless of merit and the particular problems a service must solve. Similarly, defense leaders prefer to carve up the money for new technologies into lots of small amounts and award it to many small companies to let a thousand flowers bloom, but then they seem dismayed to learn that a handful of large companies have paved over the US defense industry and only a few small flowers are struggling to survive in the cracks in the concrete.

Here, too, senior leaders must change the incentives. Rather than trying to tamp down the inherent rivalries in the defense establishment, senior leaders should unleash them. They should let stakeholders fight it out in competitions on level playing fields to demonstrate who excels at closing the kill chain, thereby transforming their rivalries from impediments to accelerants of change. Let the military services try to outdo each other. Let small start-ups try to outdo large companies, and vice versa. Let military services team up to do better things together than each could do on its own. Similarly, let novel partnerships emerge between companies large and small, old and new, hardware-focused and software-focused. Let new capabilities try to outperform legacy ones. Let everyone constantly try to outdo one another. A lot of positive outcomes could emerge more organically if senior leaders enabled all of these self-interested actors to compete to build better, faster kill chains and rewarded the winners with large contracts for their capabilities.

This is also the best way to navigate the transition from the military we have to the military we need, which will be a long and incremental process. At first, new capabilities will enhance legacy military platforms. Autonomous aircraft will make manned fighter jets more survivable and capable. Big amphibious ships may become less relevant for sending Marines onto well-defended beachheads, but creative military operators could devise important new roles for them, such as serving as mobile sea bases for unmanned underwater vehicles. Crafty military planners are still devising new ways that the B-52 bomber—after sixty-five years of service—can contribute to closing the kill chain better than other US systems, and this plotting will undoubtedly continue. These and other incremental combinations of new and old, future and present, may provide superior solutions for years to come, and military mavericks and engineers will be more likely to develop them if the incentives exist to determine what works best through real-world competitions and experiments to solve the most important operational problems.

Over time, if the future is given fair chances to compete on the merits of solving the right problems, new technologies such as intelligent machines could transition from enhancing traditional

platforms to replacing them. The same incremental experimentation will also help to define important new roles and missions for our more traditional systems. One that stands out is homeland defense. Short-range, manned fighter jets, for example, may become less relevant in a potential war against China. But those aircraft could find a vital new role closer to home as the need to strengthen our homeland defenses grows. It would not be the power projection mission these aircraft were designed for, but it would be an essential and enduring mission. Similar transitions to homeland defense or other high-priority roles could occur for a host of other legacy platforms.

The benefits of picking winners through open competitions are as much political and bureaucratic as they are military. The only way to convince our conservative and risk-averse defense establishment to adopt different kinds of military capabilities is to build these new technologies and show people what they can do under real-world conditions. Just as we must demonstrate new solutions to our competitors because of the deterrent value of doing so, we must also demonstrate them to ourselves in order to build domestic political support for new military concepts and capabilities.

Skeptical military operators and political leaders will not be convinced to give up the current capabilities that benefit them today for the promise of something better in the future that may never pan out. They will have to see it with their own eyes. They will have to see that it works and that it performs better than the systems they are buying and using today. They will have to be made to feel in a visceral way that by continuing to invest in what is familiar, they are putting their self-interest above the needs of the nation, and they will own the consequences if these legacy systems fail America's servicemembers in a future war. In short, the future will win only when political leaders become convinced the present will lose.

Restructuring the incentives for how the US government prioritizes its defense investments will have a multiplier effect among investors and engineers outside of government. I get to attend a lot of meetings, dinners, and working groups in which people are trying to bridge the divide between Washington and Silicon Valley, the defense and technology worlds, and I have come to believe that we are radically overthinking this problem. Much of the answer hinges on basic supply and demand. Again, it is a question of incentives.

On any given day, billions of dollars of private capital sit on the sidelines in America, looking for promising new ventures that could yield big returns. More of that money does not flow into the defense sector because most venture capitalists have come to believe that defense is a lousy investment, and plenty of empirical evidence supports that assumption. For decades, too many defense technologies have failed to transition from promising research and development efforts to successful military programs fielded at scale. Too many small companies doing defense work have become casualties in the "valley of death" rather than billion-dollar "unicorns." The reason there are not more success stories is not a mystery: the US government did not create the necessary incentives. It did not buy what worked best in large quantities.

There is no lack of money or will among private investors to fund a vibrant ecosystem of new defense technology companies. Investors gravitate to companies that deliver good returns on their investments. But investors do not make multi-billion-dollar bets of private money if US defense leaders do not help them pick winners. And the single most important factor in determining winners is how much revenue companies receive from the government. That is how the Department of Defense and Congress assign value to new technology: they buy a lot of it.

This is basic economics. When customers buy more of something, suppliers increase production to meet the growing demand. National defense is not a free market. The government is the only customer. But the principles are the same. If US defense leaders believe that key emerging technologies such as artificial intelligence and autonomous systems are essential for America's future military competitiveness, they have to buy those technologies at scale. No more talking a big game but failing to deliver real money. No more spreading the wealth around in lots of small slices to lots of small companies. The government has to start picking winners. It has to concentrate its investments into a few big bets—the way Eisenhower and other US leaders did in the early years of the Cold War. That is what the US government has failed to do in recent decades. That is partly why the US defense industry has consolidated and why new entrants have not broken in and grown into larger success stories. And there is no better way to identify the winners that deserve the biggest bets than through regular competitions in which they and their future-oriented solutions can compete on their merits.

If US defense leaders actually bought more of the emerging technologies and military capabilities that they say are important, private investors would have clear incentives to multiply success. The best start-ups regularly raise large infusions of private capital that are multiple times greater than the present revenue of their companies. Those private investors are not betting on what the companies are doing now but on what they could do and could become in the future, with greater resources. Most of those bets do not pan out, and the companies do not make it. But the ones that do tend to become wildly successful and make their investors very rich.

The incentives that move private capital are largely the same when it comes to the defense sector. If American leaders believe, for example, that artificial intelligence is essential for our military in the future, as they now rightly say, they should pick winners among the companies that are building these technologies and buy their capabilities in large quantities. This would lead private investors to pour exponentially more money into those companies and others in the hope of creating new moneymaking businesses. Many of those upstarts will not survive, but the ones that do could succeed tremendously. And their success will beget more success.

Prospective founders will see that it is possible to succeed in the defense market, and they might decide to build defense companies or do national security work. More of America's best engineers could be drawn to these new companies and come to see national defense as an outlet for their talents. As more companies succeed, more would be created. This is how to establish the vibrant and diversified defense industrial base that US leaders want so badly. It is not rocket science. It is mainly a question of creating the proper incentives.

This step would go a long way toward improving the relationship between the defense and technology worlds, Washington and Silicon Valley. Some American engineers will not want to work on military problems for reasons of conscience, and that is fine. At the same time, Silicon Valley is no more ideologically monolithic than Washington is. In my experience, many young engineers are open to doing defense work and would be eager to do it if more exciting opportunities were available. They are drawn to this work for many reasons, from a sense of patriotism, to a desire to succeed and distinguish themselves, to perhaps the biggest factor of all: they are engineers who are excited by the challenge of trying to solve the hardest problems. One thing our military does have is many of the hardest, most exciting problems in the world.

A big reason why many Silicon Valley engineers are frustrated with Washington is because they think America's defense leaders too often act like hypocrites. And they are not wrong. Senior leaders in the Department of Defense and Congress have a tendency to talk a big game about the importance of new technologies for the US military. But when push comes to shove, most of the biggest contracts continue to flow by the billions to legacy military platforms and the traditional defense companies that manufacture them. If Washington leaders put more of their money where their mouths are, this could entice a lot more of America's leading engineers and innovators to start working on military problems.

A change that would make it easier for Washington defense leaders to create the various new incentives that are needed is to bring back the practice of congressional earmarks. John McCain would blast me for saying this. He led the charge to ban earmarks in 2011 because he believed they had become a form of corruption, and they had. Members of Congress doled out money in nontransparent ways to programs and projects that the Department of Defense did not request and that often benefited their campaign contributors. Rather than cleaning up the earmarking process and making it fully transparent to the public, however, Congress banned the practice altogether. The unintended consequence was that the legislative branch yielded to the executive branch one of its most important prerogatives: the power to make its own independent determinations about defense programs that merit government investment. Instead, Congress relegated itself to funding only those programs that the Department of Defense deemed worthy.

The problem is that the Pentagon misses things and gets things wrong as much as the next large institution. In the past, earmarks helped members of Congress correct these errors. The Predator drone, for example, was created with funding earmarked by Congress after the Air Force consistently refused to invest in unmanned aircraft. A reformed earmarks process could play this role again by giving members of Congress a powerful tool to invest in emerging technologies for our military, especially when the Department of Defense fails to do so. The entire process would have to be transparent to prevent its corruption. The public would have to see which member of Congress was responsible for which earmarks and who benefited. This open process of accountability could encourage defensible earmarks and discourage indefensible ones.

Whether Congress brings back earmarks or not, companies that want to build a different kind of military cannot expect to win strictly by the quality of their new solutions alone. Pentagon officials regularly came through my office in the Senate, but they rarely wanted to talk about what their institutions were doing with emerging technologies or ask for additional support or money to do more or move faster. Indeed, most of the briefings I received from the Pentagon about its development of emerging technologies, and most of the meetings I had with the young companies that were doing much of this work, happened only because I asked for them. Had I not, they would have been like trees falling in the forest that I never heard.

The clear lesson that I took away is that the Americans who want to help build our future military have to lobby for themselves. And I do not mean lobby in a figurative sense. I mean the literal act of hiring lobbyists—people with intimate knowledge of and experience with the political system and budget process in Washington who can help enterprises win the attention and support of the incredibly busy leaders and staffers who can determine their fortunes in the Department of Defense and on Capitol Hill. Lobbyists may not be as all-powerful as many assume, but they can have a major impact for those trying to get their foot in the door. Indeed, the first I ever learned about SpaceX's amazing work was not because the Air Force came to tell me; it was from friends who lobbied for the company.

This also means that smaller and newer companies that want to succeed in the defense market need to play a broader political game. There is a reason why parts of the F-35 are built in every state in America, and it is not business efficiency. It is political expediency. We can bemoan that fact all we want, but it will not change. Companies that want to build future-oriented military capabilities need to create incentives that win political support for their programs. This is what the middle-sized manufacturer of the Valkyrie did. It opened a new production facility for its unmanned aircraft in Oklahoma, the home state of the current chairman of the Senate Armed Services Committee, Senator James Inhofe. Some might view this as an example of what is wrong with America's defense system. I view it as a shrewd political move that could make it more likely US troops get better technology and the future wins.

When I worked in government, I used to hear all the time that US national defense has a "cultural problem." What the critics meant by this is that the entire defense establishment, especially its acquisition processes and practices, have become so lethargic, litigious, risk averse, uncompetitive, unmeritorious, and bureaucratically calcified that it is systemically incapable of producing better outcomes. The amorphousness of the problem is what so frustrated McCain and ultimately General Milley with the Army's new pistol. Something had gone wrong, and it was hard to identify a person, process, or reason for why it happened. That is why people turned to the idea of culture.

National defense will always be fundamentally different from everything else we do in the civilian and commercial worlds. But does it have to be *this* different? Do the men and women of America's military really have to struggle *this* hard to do their jobs and get faster access to better technologies, many of which they use in their daily lives? Can't things be better?

Yes, things can be better. There is no structural or cultural reason why not. We have the money, the technological base, and the human talent. And our leaders have all of the flexibility and authorities they need, both in law and policy, to carry off the transition from the military we have to the military we need. As I have said, it ultimately comes down to incentives. If we want different and better outcomes, we have to create different and better incentives to get them. This is hardly beyond our reach. It involves doing a lot more of the commonsense things that many within our defense establishment struggle to do every day: define problems correctly and clearly, compete over the best solutions, pick winners, and spend real money on what is most important and can make our military most effective.

I know better than most just how difficult this will be, but no one should believe that it is unachievable. It can be done—but that does not necessarily mean that it will be done. Change is hard, and change of the magnitude now required will be even harder. It will require a lot more from America's senior leaders—a lot more time, imagination, resolve, and willingness to work together. And all of this ultimately depends on the answer to that one basic question that has always determined whether nations can change in the absence of catastrophe and war: Do we now believe, viscerally and actually, that there is something worse than change?

CONCLUSION

A FAILURE OF IMAGINATION

On three occasions since John McCain's death, I have lost my composure when my thoughts turned to him. The first was the night I got the call that he had passed away. The well of emotion that came over me surprised me, because we had all known for so long that this moment was coming. I thought I was prepared. What I quickly realized was that I had been so consumed with my work in the Senate trying to help his colleagues finalize the annual defense legislation that they ultimately named in McCain's honor that I had not really spent that much time thinking about McCain himself.

What came over me at that moment was the immense sense of gratitude I felt for the time I got to spend with the man. I was a part of adventures I never could have imagined—seeing parts of the world I had never seen and may never see again, meeting people at the heights and depths of human experience, being at the center of some of the most consequential national debates and international events of our time. I had been running so hard for so long that I had utterly failed to appreciate what an incredible, unforgettable, unmatchable experience this had been. I got to be a part of things that mattered, and it was because of McCain.

The second time was at his memorial service at the National Cathedral in Washington. I listened soberly to the words of presidents and other national leaders, and then I lost it halfway through the soaring rendition of the Irish ballad "Danny Boy." But as I sat there, what I found myself thinking about, beyond the many memories of McCain that played in my mind, was whether this might be a moment when things in Washington could change, when America's leaders, most of whom were sitting in that church with me, might reflect on larger questions about the unity and security of the country and perhaps conclude that we *just cannot go on like this.* And then all of the political divisiveness, mean-spiritedness, and mistrust that had been holding us back and harming our ability to prepare for the future might finally begin to fade.

What I knew, though, and what made me even more dejected, was that this would not happen. I remember thinking that this was a nice moment, but only a moment, and while many Americans and their leaders were pausing in that moment to reflect on the life of someone who represented so much of what we believe is best about our country and ourselves, we would soon be back at each other's throats—angrier, nastier, smaller, more gridlocked, and more incapable than ever of working together to do things that matter.

In less than a week, we were back to politics as usual.

This same pessimism occupied me on the third occasion when my emotions about McCain overwhelmed me. It was an overcast and unseasonably cold October morning in Annapolis, Maryland, where McCain's final resting place lies in a small cemetery on the coast of the Chesapeake Bay at the US Naval Academy. It was the first time I had been back to his grave since his death, and it did not take long for all of those old memories, emotions, and feelings of gratitude to come rushing back. But what was different this time was the overwhelming sense of sadness at the inescapable realization that things in Washington had not gotten any better since McCain's passing. Indeed, they had gotten worse—significantly, inexplicably, undeniably worse.

It is true that we now have the best opportunity in a long time to reimagine our national defense, for many people within our defense establishment seem more motivated to prepare our nation for a future international security environment that new threats and new technology are transforming. Some good things are happening. And this a testament to how many strategic advantages we Americans still have going for us. We have so many decent, hardworking, dedicated people. We have such amazing technology in our country. We have all the money we need. The bigger problem, however, remains: we just cannot get out of our own way.

I know things can be different, and I want them to be different every bit as much as when I was sitting in the National Cathedral. But it is hard to imagine, much less truly *believe,* that things will change when I see what is happening. When, for instance, the president does things nearly every day that further divide an already divided country and diminish our standing in the world. It is hard to believe when Congress fails yet again to provide our military with a timely budget, when our defense establishment continues to spend billions of dollars on the same things that continue to fail, when Americans in our military too often struggle to get access to the best technology that our nation has to offer for reasons of politics and bureaucracy. It is especially hard to believe when, facing what is quickly becoming the most significant national security challenge in our history—a technologically advanced peer competitor in China that is nationally mobilized and moving rapidly to supersede us and become the world's preeminent power—America collectively, tragically, seems to be doing more to make our competitors' jobs easier for them.

Would these kinds of self-defeating behaviors be happening if America was actually serious about the growing threats we face? And yet, they *are* happening, and because they are, it is hard to escape the conclusion that America still is not serious—that most of the underlying problems that have brought about our current national defense crisis still exist. None of these problems needs to persist and plague us. We have every advantage to address them. So how do we explain the fact that we just seem to be so much less than the challenges we face? The problem seems to be what it has been all along—a failure of imagination.

I understand how hard it is, politically and bureaucratically, to make the kinds of changes to America's national defense that I believe are necessary, especially in our aberrational and highly dysfunctional political environment. But there are worse things—much worse things—than the political pain associated with making those changes. And it is the inability or unwillingness of a critical mass of our leaders in Washington to appreciate this in a real and visceral way that represents our deeper failure of imagination. If Americans still cannot imagine something worse than change, then talk of change will abound, but actual change will continue to elude us.

There is still a pervasive belief in Washington, largely the result of the exceptional period of history from which we are emerging, that somehow everything will work out well for America. We cannot envision it otherwise because our old position of unrivaled dominance casts such a long shadow over our imagination. If that view continues to persist, things will not change fast enough. And we should be under no illusion of the risks we are running.

Our failure to adapt will not stop others from doing so. If America does not change itself, change will still happen. A new revolution in military affairs will still unfold, but it will primarily benefit others. And when revolutionary change does arrive, it will arrive not as part of a plan that Washington has led and controlled but rather as something that happens to us—such as the breakdown of deterrence and the loss of a war—something we could have prevented, that we could have shaped for our benefit, that we should have been able to imagine. At that point, we would change because we had to, but by then it would be too late. We would have lost much of our ability to control our own destiny. We would be living at the mercy of our rivals. The damage would be done.

None of this is inevitable. There is still reason for hope. But as McCain was fond of saying, hope is not a strategy. The responsibility for defending America lies with us, and time is running out.

ACKNOWLEDGMENTS

This book is the culmination of a long journey—an arduous journey, to be sure, but one made more joyful and exhilarating because it has fundamentally been a process of learning in the company of friends. All of the ideas contained in this book can be traced back to conversations, debates, and a few arguments that I have had the good fortune of sharing over many years with friends and teammates, especially during my time working in the US government.

None of this would have been possible without the friendship and mentorship of Senator John McCain, who took a chance on me and brought me into his confidence for nearly a decade. It is often said that no man is a hero to his valet, but there was not a day that went by that I did not know and appreciate the fact that I got to work for a genuine hero. I was surrounded for all those years by other decent and dedicated public servants, both in McCain's Senate office and on the staff of the Senate Armed Services Committee. I learned and benefited from my friendship and collaboration with all of them (as well as some shared suffering). It was the adventure of a lifetime, and I treasure every day that I spent in public service—well, most of them anyway.

This book began as a paper that Nick Burns, Joe Nye, and Condoleezza Rice asked me to write and present to the Aspen Strategy Group in August 2018. I am grateful to them for giving me that opportunity and to our fellow Strategy Group members for their thoughtful and helpful feedback on the ideas that became this book.

I owe special thanks to Bill Burns, president of the Carnegie Endowment for International Peace, for believing in this project and providing a valuable home for me to pursue it at Carnegie. Bill has been a friend and mentor for more than a decade, and his thoughtful feedback on early drafts of this book improved it immeasurably. I also appreciate the support and assistance I have received from Tom Carothers, Matan Chorev, Evan Feigenbaum, Jen Psaki, Doug Farrar, and the rest of my Carnegie colleagues. And I am especially grateful to Ashley Tellis and Salman Ahmed for reading a draft of this book and providing valuable feedback.

Other friends also read large sections of this book and provided critical reactions that improved it immensely: Zach Mears, Ryan Evans, Diem Salmon, Matt Waxman, Paul Scharre, Dan Patt, Scott Cuomo, Jason Matheny, Andrew May, Tom Mahnken, Truman Anderson, and Pauline Shanks Kaurin. They have consistently helped me to formulate and work out the ideas within these pages, as have a great number of other friends, especially Bridge Colby, Mara Karlin, Clint Hinote, John Richardson, Christian Wortman, Bob Work, Roger Zakheim, Kori Schake, Vance Serchuk, Andrew Krepinevich, Mike Gallagher, Jim Baker, Kathleen Hicks, Mark Montgomery, James Hickey, Dave Ochmanek, Bryan Clark, Mike Brown, Fred Kennedy, Michael Horowitz, Mackenzie Eaglen, and Rush Doshi.

This book has also represented the beginning of a new journey since I have left the Senate and joined Anduril Industries. I am grateful to Anduril's founders—Palmer Luckey, Trae Stephens, Brian Schimpf, Matt Grimm, and Joe Chen—for allowing me to be part of their phenomenal team. They and our many other colleagues have taught me an immense amount over the past year about technology, engineering, business, and investment. I am especially grateful to Matt Steckman and Eliot Pence, who, in addition to being outstanding partners in our daily work together, read early versions of this book and provided valuable feedback.

I would like to thank my agent, Andrew Wylie, and his team at the Wylie Agency for their tenacious work on behalf of this project and me. I am grateful to David Lamb, my editor at Hachette Books, who has been a valuable partner in this endeavor and who has improved it immensely, as well as Christina Palaia and the rest of the Hachette team who have helped to bring this book into existence: Amber Morris, Mandy Kain, Mike Giarratano, and Michael Barrs. I also want to thank Paul Whitlach, who first took a chance on this book and on me and helpfully guided my initial work.

I owe an immeasurable debt of gratitude to my parents, Eric and Christine, who have always selflessly poured their love and support into my brother, Dieter, and me, and who clearly started me on my journey as a writer through their many years of committed service as my editors, proofreaders, and leading critics, including on this book. I owe them everything.

I want to thank my children, Sam and Oliver, for their patience with my absence during so many long days of writing. They were nothing but understanding, encouraging, and supportive of me throughout this entire project. Most of all, I am grateful to my wife, Molly, the love of my life and my partner in all things. No one did more than she did to help and support me, advise and counsel me, sustain me and improve my thinking, bear burdens large and small, and generally hold the life of our family together throughout my work on this project. This book is dedicated to Molly because it simply would not have been possible without her.

NOTES

INTRODUCTION: PLAYING A LOSING GAME

1. General Joseph Dunford, Testimony Before the Senate Armed Services Committee, June 13, 2017, https://www.armed-services.senate.gov/hearings/17-06-13-department-of-defense-budget-posture.
2. David Ochmanek, Peter Wilson, Brenna Allen, John Speed Meyers, and Carter C. Price, *U.S. Military Capabilities and Forces for a Dangerous World* (Santa Monica, CA: RAND Corporation, 2017), xii.
3. Eric Edelman and Gary Roughhead, *Providing for the Common Defense: The Assessment and Recommendations of the National Defense Strategy Commission* (Washington, DC: United States Institute of Peace, 2018), https://www.usip.org/sites/default/files/2018-11/providing-for-the-common-defense.pdf.
4. Remarks of David Ochmanek at the Center for a New American Security, Panel Discussion: A New American Way of War, May 7, 2019, https://www.cnas.org/events/panel-discussion-a-new-american-way-of-war.
5. Andrew J. Nathan, "The Chinese World Order," *New York Review of Books*, October 12, 2017, https://www.nybooks.com/articles/2017/10/12/chinese-world-order/.
6. Callum Paton, "World's Largest Economy in 2030 Will Be China, Followed by India, with the U.S. Dropping to Third," *Newsweek*, January 10, 2019, https://www.newsweek.com/worlds-largest-economy-2030-will-be-china-followed-india-us-pushed-third-1286525.

CHAPTER 1: WHAT HAPPENED TO YODA'S REVOLUTION

1. For more about Marshall, see Andrew Krepinevich and Barry Watts, *The Last Warrior: Andrew Marshall and the Shaping of Modern American Defense Strategy* (New York: Basic Books, 2015).
2. Thomas A. Keaney and Eliot A. Cohen, *Gulf War Air Power Survey Summary Report* (Washington, DC: Department of the Air Force, 1993), 251.
3. Keaney and Cohen, *Gulf War Air Power Survey,* 237.
4. William A. Owens, with Ed Offley, *Lifting the Fog of War* (Baltimore: Johns Hopkins University Press, 2001), 14.
5. Thomas G. Mahnken, "Weapons: The Growth & Spread of the Precision-Strike Regime," *Daedalus* 140, no. 3 (2011): 45–57.
6. US Department of Defense, *Report of the Quadrennial Defense Review* (Washington, DC: Department of Defense, May 1997), 41.
7. Michael Vickers and Robert Martinage, *Future Warfare 20XX Wargame Series: Lessons Learned Report* (Washington, DC: Center for Strategic and Budgetary Assessments, December 2001), 1.
8. Vickers and Martinage, *Future Warfare 20XX Wargame Series,* 6.
9. Todd Harrison, *Defense Modernization Plans Through the 2020s: Addressing the Bow Wave* (Washington, DC: Center for Strategic and International Studies, 2016), 6.
10. Tyler Rogoway, "The Alarming Case of the USAF's Mysteriously Missing Unmanned Combat Air Vehicles," *The Drive,* June 9, 2016, https://www.thedrive.com/the-war-zone/3889/the-alarming-case-of-the-usafs-mysteriously-missing-unmanned-combat-air-vehicles.

CHAPTER 2: LITTLE GREEN MEN AND ASSASSIN'S MACE

1. David A. Shlapak and Michael Johnson, *Reinforcing Deterrence on NATO's Eastern Flank: Wargaming the Defense of the Baltics* (Santa Monica, CA: RAND Corporation, 2016), https://www.rand.org/pubs/research_reports/RR1253.html.
2. Quoted in Molly McKew, "The Gerasimov Doctrine," *Politico Magazine,* September/October 2017, https://www.politico.com/magazine/story/2017/09/05/gerasimov-doctrine-russia-foreign-policy-215538.
3. Kurt Campbell and Ely Ratner, "The China Reckoning: How Beijing Defied American Expectations," *Foreign Affairs,* March

/April 2018, https://www.foreignaffairs.com/articles/china/2018-02-13/china-reckoning.

4. Jeffrey Engstrom, *Systems Confrontation and Systems Destruction Warfare* (Santa Monica, CA: RAND Corporation, 2018), 10.
5. Josh Rogin, "NSA Chief: Cybercrime Constitutes the 'Greatest Transfer of Wealth in History,'" *Foreign Policy*, July 9, 2012, https://foreignpolicy.com/2012/07/09/nsa-chief-cybercrime-constitutes-the-greatest-transfer-of-wealth-in-history/.
6. Rush Doshi, "Hu's to Blame for China's Foreign Assertiveness," Brookings Institution, January 22, 2019, https://www.brookings.edu/articles/hus-to-blame-for-chinas-foreign-assertiveness/.
7. *The 9/11 Commission Report: Final Report of the National Commission on Terrorist Attacks upon the United States* (New York: W. W. Norton, 2004), 362.
8. Bob Work, "The Third Offset Strategy" (speech at Reagan Defense Forum, Reagan Presidential Library, Simi Valley, CA, November 7, 2015), https://dod.defense.gov/News/Speeches/Speech-View/Article/628246/reagan-defense-forum-the-third-offset-strategy/.

CHAPTER 3: A TALE OF TWO CITIES

1. Dwight D. Eisenhower, "Scientific and Technological Resources as Military Assets" (memorandum for Directors and Chiefs of War Department General and Special Staff Divisions and Bureaus and the Commanding Generals of the Major Commands, Office of the Chief of Staff, War Department, Washington, DC, April 30, 1946).
2. Dwight Eisenhower, notes for address to the Industrial Associations, Chicago, 1947, Eisenhower Presidential Library, https://www.eisenhower.archives.gov/all_about_ike/speeches.html, quoted in Robert O. Work and Greg Grant, *Beating the Americans at Their Own Game: An Offset Strategy with Chinese Characteristics* (Washington, DC: Center for a New American Security, 2019), 2.
3. For a fuller discussion of Eisenhower and Schriever, see Neil Sheehan, *A Firey Peace in a Cold War* (New York: Vintage Books, 2009).
4. Margaret O'Mara, "Silicon Valley Can't Escape the Business of War," *New York Times*, October 26, 2018, https://www.nytimes.com/2018/10/26/opinion/amazon-bezos-pentagon-hq2.html.

5. Pierre A. Chao, "The Structure and Dynamics of the Defense Industry" (remarks at the Security Studies Program Seminar, Center for Security and International Studies, Washington, DC, March 2, 2005), http://web.mit.edu/SSP/seminars/wed_archives05spring/chao.htm.
6. Andrew P. Hunter, Samantha Cohen, Gregory Sanders, Samuel Mooney, and Marielle Roth, *New Entrants and Small Business Graduation in the Market for Federal Contracts* (Washington, DC: Center for Strategic and International Studies, 2018), VIII, https://csis-prod.s3.amazonaws.com/s3fs-public/publication/181120_NewEntrantsandSmallBusiness_WEB.pdf?GoT2hzpdiSBJXUyX.lMMoHHerBrzzoEf.
7. Joe Gould, "American Exodus? 17,000 US Defense Suppliers May Have Left the Defense Sector," *Defense News*, December 14, 2017, https://www.defensenews.com/breaking-news/2017/12/14/american-exodus-17000-us-defense-suppliers-may-have-left-the-defense-sector/.

CHAPTER 4: INFORMATION REVOLUTION 2.0

1. Lockheed Martin, "Multi-Mission Capability for Emerging Global Threats," F-35 Lightning II Lockheed Martin, https://www.f35.com/about/capabilities.
2. Rob Csongor, "Tesla Raises the Bar for Self-Driving Cars," Nvidia, April 23, 2019, https://blogs.nvidia.com/blog/2019/04/23/tesla-self-driving/.
3. Colin Clark, "Cardillo: 1 Million Times More GEOINT Data in 5 Years," *Breaking Defense*, June 5, 2017, https://breakingdefense.com/2017/06/cardillo-1-million-times-more-geoint-data-in-5-years/.
4. Bernard Marr, "How Much Data Do We Create Every Day? The Mind-Blowing Stats That Everyone Should Read," *Forbes*, May 21, 2018, https://www.forbes.com/sites/bernardmarr/2018/05/21/how-much-data-do-we-create-every-day-the-mind-blowing-stats-everyone-should-read/#20aa810860ba.
5. Jason Metheny, "Four Emerging Technologies and National Security," in *Technology and National Security: Maintaining America's Edge*, ed. Leah Bitounis and Jonathon Price (Washington, DC: Aspen Institute, 2019), 33.
6. Stephen Shankland, "Elon Musk Says Neuralink Plans 2020 Human Test of Brain-Computer Interface," CNET, July 17, 2019, https://www.cnet.com/news/elon-musk-neuralink-works-monkeys-human-test-brain-computer-interface-in-2020/.

CHAPTER 5: SOMETHING WORSE THAN CHANGE

1. Norman Dixon, *The Psychology of Military Incompetence* (New York: Basic Books, 1976), 159.
2. Geoffrey Till, "Adopting the Aircraft Carrier: The British, American, and Japanese Cases," in *Military Innovation in the Interwar Period,* ed. Williamson Murray and Allan R. Millet (New York: Cambridge University Press, 1996), 210. See also Stephen Peter Rosen, *Winning the Next War: Innovation and the Modern Military* (Ithaca, NY: Cornell University Press, 1991).
3. Barry Posen, *The Sources of Military Doctrine: France, Britain, and Germany Between the World Wars* (Ithaca, NY: Cornell University Press, 1984).
4. William Owens, with Ed Offley, *Lifting the Fog of War* (Baltimore: Johns Hopkins University Press, 2001), 20.
5. Robert O. Work and Greg Grant, *Beating the Americans at Their Own Game: An Offset Strategy with Chinese Characteristics* (Washington, DC: Center for a New American Security, 2019), 1.
6. Graham Allison, *Destined for War: Can America and China Escape Thucydides's Trap?* (New York: Houghton Mifflin Harcourt, 2017).
7. Jude Blanchette, *China's New Red Guards: The Return of Radicalism and the Rebirth of Mao Zedong* (New York: Oxford University Press, 2019).
8. "Document 9: A ChinaFile Translation," ChinaFile, November 8, 2013, http://www.chinafile.com/document-9-chinafile-translation.
9. Bill Bishop, "Engineers of the Soul: Ideology in Xi Jinping's China by John Garnaut," Sinocism, January 16, 2019, https://sinocism.com/p/engineers-of-the-soul-ideology-in.
10. For a more detailed discussion of the Chinese government's views of technology and security, see the critical and extensive work done by Elsa Kania, including *Battlefield Singularity: Artificial Intelligence, Military Revolution, and China's Future Military Power* (Washington, DC: Center for a New American Security, November 28, 2017).
11. Patrick Tucker, "China Will Surpass US in AI Around 2025, Says Google's Eric Schmidt," DefenseOne, https://www.defenseone.com/technology/2017/11/google-chief-china-will-surpass-us-ai-around-2025/142214/.
12. Kania, *Battlefield Singularity,* 15.

13. David Lague and Benjamin Kang Lim, "How China Is Replacing America as Asia's Military Titan," Reuters, April 23, 2019, https://www.reuters.com/investigates/special-report/china-army-xi/.
14. Joe McDonald, "Military Parade Will Be Rare Look at China's Arms, Ambitions," Associated Press, September 29, 2019, https://www.apnews.com/ca6f789c8b95493ba0a3327a042da596.
15. David Lague and Benjamin Kang Lim, "China's Vast Fleet Is Tipping the Balance in the Pacific," Reuters, April 30, 2019, https://www.reuters.com/investigates/special-report/china-army-navy.
16. Captain James Fanell (USN, Ret.), *China's Global Naval Strategy and Expanding Force Structure: Pathway to Hegemony* (testimony before the House Permanent Select Committee on Intelligence, hearing on China's Worldwide Military Expansion, May 17, 2018), https://docs.house.gov/meetings/IG/IG00/20180517/108298/HHRG-115-IG00-Wstate-FanellJ-20180517.pdf.

CHAPTER 6: A DIFFERENT KIND OF ARMS RACE

1. Department of Defense, *Directive 3000.09: Autonomy in Weapon Systems* (Washington, DC: Department of Defense, November 21, 2012, updated May 8, 2017), https://www.esd.whs.mil/portals/54/documents/dd/issuances/dodd/300009p.pdf.
2. Elsa Kania, "China's Strategic Ambiguity and Shifting Approach to Lethal Autonomous Weapons Systems," Lawfare, April 17, 2018, https://www.lawfareblog.com/chinas-strategic-ambiguity-and-shifting-approach-lethal-autonomous-weapons-systems.
3. Kania, *Battlefield Singularity*, 22.
4. Kania, *Battlefield Singularity*, 23.
5. Kyle Mizokami, "Here's What We Saw at China's Gigantic Military Parade," Foxtrot Alpha, October 3, 2019, https://foxtrotalpha.jalopnik.com/heres-what-we-saw-at-china-s-gigantic-military-parade-1838676610.
6. Billy Perrigo, "A Global Arms Race for Killer Robots Is Transforming the Battlefield," *Time*, April 9, 2018, https://time.com/5230567/killer-robots/.
7. Mark Clayton, "The New Cyber Arms Race," *Christian Science Monitor*, March 7, 2011, https://www.csmonitor.com/USA/Military/2011/0307/The-new-cyber-arms-race.

8. Peter Apps, "Are China, Russia Winning the AI Arms Race?" Reuters, January 15, 2019, https://www.reuters.com/article/us-apps-ai-commentary/commentary-are-china-russia-winning-the-ai-arms-race-idUSKCN1P91NM.
9. Richard M. Harrison, "Welcome to the Hypersonic Arms Race," The Buzz (blog), National Interest, January 19, 2019, https://nationalinterest.org/blog/buzz/welcome-hypersonic-arms-race-42002.
10. Andrew Ross Sorkin, Stephen Grocer, Tiffany Hsu, and Gregory Schmidt, "5G Is the New Arms Race with China," *New York Times,* January 28, 2019, https://www.nytimes.com/2019/01/28/business/dealbook/us-china-5g-huawei-internet.html.
11. Martin Giles, "The US and China Are in a Quantum Arms Race That Will Transform Warfare," *MIT Technology Review,* January 3, 2019, https://www.technologyreview.com/s/612421/us-china-quantum-arms-race/.
12. Sy Mukherjee, "Goldman Sachs: China Is Beating the U.S. in the Gene Editing Arms Race," *Fortune,* April 23, 2018, https://fortune.com/2018/04/13/goldman-sachs-china-gene-editing-race/.
13. Lara Seligman, "The New Space Race," *Foreign Policy,* May 14, 2019, https://foreignpolicy.com/2019/05/14/the-new-space-race-china-russia-nasa/.
14. See, for example: Justin Sherman, "Reframing the U.S.-China AI Arms Race," *New America,* March 6, 2019, https://www.newamerica.org/cybersecurity-initiative/reports/essay-reframing-the-us-china-ai-arms-race; Michael Horowitz, "The Algorithms of August," *Foreign Policy,* September 12, 2018, https://foreignpolicy.com/2018/09/12/will-the-united-states-lose-the-artificial-intelligence-arms-race; Paul Scharre, "The Real Dangers of an AI Arms Race," *Foreign Affairs,* May/June 2019, https://www.foreignaffairs.com/articles/2019-04-16/killer-apps.
15. Michael Horowitz, "Artificial Intelligence, International Competition, and the Balance of Power," *Texas National Security Review* 1, no. 3 (May 2018): 41.
16. Sophia Chen, "Why Google's Quantum Victory Is a Huge Deal—and a Letdown," *Wired,* September 26, 2019, https://www.wired.com/story/why-googles-quantum-computing-victory-is-a-huge-deal-and-a-letdown/.

17. Edwin Pedault, John Gunnels, Dmitri Maslov, and Jay Gambetta, "On 'Quantum Supremacy,'" IBM Research Blog (blog), IBM, October 21, 2019, https://www.ibm.com/blogs/research/2019/10/on-quantum-supremacy.
18. Stephanie Nebehay, "U.N. Says It Has Credible Reports That China Holds Millions of Uighurs in Secret Camps," Reuters, August 10, 2018, https://www.reuters.com/article/us-china-rights-un/u-n-says-it-has-credible-reports-that-china-holds-million-uighurs-in-secret-camps-idUSKBN1KV1SU.
19. Josh Gabbatiss, "World's First Genetically Altered Babies Born in China, Scientist Claims," *Independent,* November 26, 2018, https://www.independent.co.uk/news/world/asia/china-babies-genetically-edited-altered-twins-scientist-dna-crispr-a8651536.html.
20. See Kai-Fu Lee, *AI Superpowers: China, Silicon Valley, and the New World Order* (New York: Houghton Mifflin Harcourt, 2018).
21. Gregory Allen, *Understanding China's AI Strategy: Clues to Chinese Strategic Thinking on Artificial Intelligence and National Security* (Washington, DC: Center for a New American Security, 2019), 19.
22. Kate O'Keeffe and Jeremy Page, "China Taps Its Private Sector to Boost Its Military, Raising Alarms," *Wall Street Journal,* September 25, 2019, https://www.wsj.com/articles/china-taps-its-private-sector-to-boost-its-military-raising-alarms-11569403806.

CHAPTER 7: HUMAN COMMAND, MACHINE CONTROL

1. See further, Paul Scharre, *Army of None: Autonomous Weapons and the Future of War* (New York: W. W. Norton, 2018); Heather Roff with David Danks, "The Necessity and Limits of Trust in Autonomous Weapons System," *Journal of Military Ethics,* 2018; Joe Chapa, "Drone Ethics and the Civil-Military Gap," *War on the Rocks,* June 28, 2017, https://warontherocks.com/2017/06/drone-ethics-and-the-civil-military-gap; Julia MacDonald and Jacquelyn Schneider, "Trust, Confidence, and the Future of War," *War on the Rocks,* February 5, 2018; Ronald Arkin, "Lethal Autonomous Systems and the Plight of the Non-combatant," in *The Political Economy of Robots,* ed. Ryan Kiggins (London: Palgrave Macmillan, 2017); and Department of Defense, Defense Innovation Board, *AI Principles: Recommendations of the Ethical Use of Artificial Intelligence by the Department of Defense* (Washington, DC: Department of Defense, 2019), https://

media.defense.gov/2019/Oct/31/2002204458/-1/-1/0/DIB_AI_PRINCIPLES_PRIMARY_DOCUMENT.PDF.

2. Bryan Clark, Daniel Patt, and Harrison Schramm, "Decision Maneuver: The Next Revolution in Military Affairs," *Over the Horizon,* April 29, 2019, https://othjournal.com/2019/04/29/decision-maneuver-the-next-revolution-in-military-affairs/.
3. James Vincent, "This Is When AI's Top Researchers Think Artificial General Intelligence Will Be Achieved," *The Verge,* November 27, 2018, https://www.theverge.com/2018/11/27/18114362/ai-artificial-general-intelligence-when-achieved-martin-ford-book.
4. See Kenneth Anderson, Daniel Reisner, and Matthew Waxman, "Adapting the Law of Armed Conflict to Autonomous Weapon Systems," *International Law Studies,* 90:386 (2014), 398–401.
5. Anderson, Reisner, and Waxman, "Adapting the Law of Armed Conflict," 401–402.
6. Darrell M. West, "Brookings Survey Finds Divided Views on Artificial Intelligence for Warfare, but Support Rises If Adversaries Are Developing It," Brookings Institution, August 29, 2018, https://www.brookings.edu/blog/techtank/2018/08/29/brookings-survey-finds-divided-views-on-artificial-intelligence-for-warfare-but-support-rises-if-adversaries-are-developing-it/.

CHAPTER 8: A MILITARY INTERNET OF THINGS

1. Tyler Rogoway, "More Details Emerge on Kratos' Optionally Expendable Air Combat Drones," The Warzone, The Drive, February 7, 2017, thedrive.com/the-war-zone/7449/more-details-on-kratos-optionally-expendable-air-combat-drones-emerge?iid=sr-link1.
2. Peter Rathmell, "Navy, Boeing Partner to Build Deep-Sea Drone," *Navy Times,* June 12, 2017, https://www.navytimes.com/news/your-navy/2017/06/12/navy-boeing-partner-to-build-deep-sea-drone.
3. Lockheed Martin, "Building the F-35: Combining Teamwork and Technology," F-35 Lightning II Lockheed Martin, https://www.f35.com/about/life-cycle/production.

CHAPTER 9: MOVE, SHOOT, COMMUNICATE

1. Jan Bloch, *The Future of War* (Boston: World Peace Foundation, 1898), xxvii.

2. Robert Scales, "The Great Duality and the Future of the Army: Does Technology Favor the Offensive or Defensive?" *War on the Rocks,* September 3, 2019, https://warontherocks.com/2019/09/the-great-duality-and-the-future-of-the-army-does-technology-favor-the-offensive-or-defensive.
3. The role of open source information was critical to the findings of the Joint Investigative Team, an international body established to investigate the shootdown of Malaysian Airlines Flight 17. The team summarized its findings in a press conference and a video released on May 28, 2018, which can be viewed at https://youtu.be/rhyd875Qtlg.
4. For a comprehensive overview of China's land reclamation efforts and the role of open source information in documenting it, see the critical work done by the Asia Maritime Transparency Initiative at the Center for Strategic and International Studies, https://amti.csis.org/accounting-chinas-deployments-spratly-islands.
5. Sebastian Sprenger, "A German Radar Maker Says It Tracked the F-35 Stealth Fighter in 2018—from a Pony Farm," *Business Insider,* September 30, 2019, https://www.businessinsider.com/german-radar-maker-hensoldt-says-it-tracked-f35-in-2018-2019-9.
6. United States Government Accountability Office, *Defense Acquisitions: Assessments of Selected Weapons Programs*, Report to Congressional Committees (Washington, DC: US Government Accountability Office, 2014), 115, https://www.gao.gov/assets/670/662184.pdf.
7. Patrick Tucker, "It's Now Possible to Telepathically Communicate with a Drone Swarm," *Defense One,* September 6, 2018, https://www.defenseone.com/technology/2018/09/its-now-possible-telepathically-communicate-drone-swarm/151068/.

CHAPTER 10: DEFENSE WITHOUT DOMINANCE

1. United States Department of Defense, *Summary of the 2018 National Defense Strategy of the United States of America* (Washington, DC: Department of Defense, 2018), https://dod.defense.gov/Portals/1/Documents/pubs/2018-National-Defense-Strategy-Summary.pdf.
2. For further reading, see, for example: Elbridge Colby, "How to Win America's Next War," *Foreign Policy,* Spring 2019; Elbridge Colby, Testimony Before the Senate Armed Services Committee,

January 29, 2019, https://www.armed-services.senate.gov/download/colby_01-29-19; T. X. Hammes, *The Melians Revenge*, Atlantic Council Issue Brief (Washington, DC: Atlantic Council, 2019); Mara Karlin, "How to Read the 2018 National Defense Strategy," Brookings Institution, January 21, 2018, https://www.brookings.edu/blog/order-from-chaos/2018/01/21/how-to-read-the-2018-national-defense-strategy; Lt. Gen. David A. Deptula, Heather Penney, Maj. Gen. Lawrence A. Stutzriem, and Mark Gunzinger, *Restoring America's Military Competitiveness: Mosaic Warfare* (Arlington, VA: Mitchell Institute for Aerospace Studies, 2019); and Mackenzie Eaglen, "Just Say No: The Pentagon Needs to Drop the Distractions and Move Great Power Competition Beyond Lip Service," *War on the Rocks*, October 28, 2019, https://warontherocks.com/2019/10/just-say-no-the-pentagon-needs-to-drop-the-distractions-and-move-great-power-competition-beyond-lip-service.

3. Christopher M. Dougherty, *Why America Needs a New Way of War* (Washington, DC: Center for a New American Security, 2019), https://s3.amazonaws.com/files.cnas.org/CNAS+Report+-+ANAWOW+-+FINAL2.pdf.
4. See, for example: Kathleen Hicks and Alice Hunt Friend, *By Other Means: Part 1: Campaigning in the Gray Zone* (Washington, DC: Center for Strategic and International Studies, 2019), https://csis-prod.s3.amazonaws.com/s3fs-public/publication/Hicks_GrayZone_interior_v4_FULL_WEB_0.pdf; and Melissa Dalton, Kathleen Hicks, Alice Hunt Friend, Lindsey Shepherd, and Joseph Federici, *By Other Means: Part 2: Adapting to Compete in the Gray Zone* (Washington, DC: Center for Strategic and International Studies, 2019), https://csis-prod.s3.amazonaws.com/s3fs-public/publication/Hicks_GrayZone_II_interior_v8_PAGES.pdf.
5. Anne Gearan, Damian Paletta, and John Wagner, "Trump Takes Aim at Foreign Leaders and Critics Before Heading to Economic Summit in Japan," *Washington Post*, June 26, 2019, https://www.washingtonpost.com/politics/as-he-heads-to-japan-trump-complains-of-lopsided-military-obligations/2019/06/26/efaa5870-97fb-11e9-830a-21b9b36b64ad_story.html?noredirect=on.

CHAPTER 11: BUREAUCRACY DOES ITS THING

1. Kyle Jahner, "Army Chief: You Want a New Pistol? Send Me to Cabela's with $17 Million," *Army Times*, March 27, 2016, https://www.armytimes.com/news/your-army/2016/03/28/army-chief-you-want-a-new-pistol-send-me-to-cabela-s-with-17-million/.
2. John Dowdy and Chandru Krishnamurthy, "Defense in the 21st Century: How Artificial Intelligence Might Change the Character of Conflict," in *Technology and National Security: Maintaining America's Edge*, ed. Leah Bitounis and Jonathon Price (Washington, DC: Aspen Institute, 2019), 83.
3. Scott Conn, Testimony Before the Seapower Subcommittee, in *Statement of the Honorable James F. Geurts Assistant Secretary of the Navy for Research, Development and Acquisition ASN (RD&A), and Lieutenant General Steven Rudder Deputy Commandant for Aviation, and Rear Admiral Scott Conn Director Air Warfare Before the Seapower Subcommittee of the Senate Armed Services Committee on Department of the Navy Aviation Programs*, April 10, 2019, https://www.armed-services.senate.gov/hearings/19-04-10-marine-corps-ground-modernization-and-naval-aviation-programs.

CHAPTER 12: HOW THE FUTURE CAN WIN

1. Jen Judson, "Army's 'Night Court' Finds $25 Million to Reinvest in Modernization Priorities," *Defense News*, October 8, 2018, https://www.defensenews.com/digital-show-dailies/ausa/2018/10/08/armys-night-court-finds-25-billion-to-reinvest-in-modernization-priorities/.
2. General David H. Berger, *Commandant's Planning Guidance* (Washington, DC: United States Marine Corps, 2019), https://www.hqmc.marines.mil/Portals/142/Docs/%2038th%20Commandant%27s%20Planning%20Guidance_2019.pdf?ver=2019-07-16-200152-700.
3. Sydney J. Freedberg Jr., "US 'Gets Its Ass Handed to It' in Wargames: Here's a $24 Billion Fix," *Breaking Defense*, March 7, 2019, https://breakingdefense.com/2019/03/us-gets-its-ass-handed-to-it-in-wargames-heres-a-24-billion-fi.

BIBLIOGRAPHY

Allen, Gregory. *Understanding China's AI Strategy: Clues to Chinese Strategic Thinking on Artificial Intelligence and National Security* (Washington, DC: Center for a New American Security, 2019).

Allison, Graham. *Destined for War: Can America and China Escape Thucydides's Trap?* (New York: Houghton Mifflin Harcourt, 2017).

Anderson, Cory T., David Blair, Mike Byrnes, Joe Chapa, Amanda Callazzo, Scott Cuomo, Olivia Garard, Ariel M. Schuetz, and Scott Vanoort. "Trust, Troops, and Reapers: Getting 'Drone' Research Right." *War on the Rocks,* April 3, 2018, https://warontherocks.com/2018/04/trust-troops-and-reapers-getting-drone-research-right.

Anderson, Kenneth, with Daniel Reisner and Matthew Waxman. "Adapting the Law of Armed Conflict to Autonomous Weapon Systems." *International Law Studies* 90:386 (2014).

Anderson, Kenneth, and Matthew C. Waxman. "Law and Ethics for Autonomous Weapon Systems: Why a Ban Won't Work and How the Laws of War Can." Jean Perkins Task Force on National Security and Law Essay Series, WCL Research Paper 2013-11, Columbia Public Law Research Paper 13-351, Stanford University, Hoover Institution, American University, April 10, 2013.

Arkin, Ronald. "The Case for Ethical Autonomy in Unmanned Systems." *Journal of Military Ethics* 9, no. 4 (2010): 332–341.

———. "Lethal Autonomous Systems and the Plight of the Non-combatant." In *The Political Economy of Robots,* edited by Ryan Kiggins (London: Palgrave Macmillan, 2017).

Berger, David H. *Commandant's Planning Guidance* (Washington, DC: United States Marine Corps, 2019). https://www.hqmc.marines.mil

/Portals/142/Docs/%2038th%20Commandant%27s%20Planning%20Guidance_2019.pdf?ver=2019-07-16-200152-700.

Bitounis, Leah, and Jonathon Price, eds. *Technology and National Security: Maintaining America's Edge* (Washington, DC: The Aspen Institute, 2019).

Blair, Dennis, and Robert D. Atkinson. "Overcoming the China Challenge." *American Interest* 14, no. 4 (2018).

Blanchette, Jude. *China's New Red Guards: The Return of Radicalism and the Rebirth of Mao Zedong* (New York: Oxford University Press, 2019).

Bloch, Jan. *The Future of War* (Boston: World Peace Foundation, 1898).

Bostrom, Nick. *Superintelligence: Paths, Dangers, Strategies* (Oxford: Oxford University Press, 2014).

Brands, Hal, and Zack Cooper. "After the Responsible Stakeholder, What? Debating America's China Strategy." *Texas National Security Review* 2, no. 2 (February 2019).

Brecher, Joseph, Heath Niemi, and Andrew Hill. "My Droneski Just Ate Your Ethics," *War on the Rocks*, August 10, 2016, https://warontherocks.com/2016/08/my-droneski-just-ate-your-ethics.

Brunstetter, Daniel, and Megan Braun. "The Implications of Drones on the Just War Tradition." *Ethics and International Affairs* 25, no. 3 (Fall 2011).

Campbell, Kurt, and Ely Ratner. "The China Reckoning: How Beijing Defied American Expectations." *Foreign Affairs*, March/April 2018.

Chapa, Joe. "Drone Ethics and the Civil-Military Gap." *War on the Rocks*, June 28, 2017, https://warontherocks.com/2017/06/drone-ethics-and-the-civil-military-gap.

———. "The Sunset of the Predator: Reflections on the End of an Era." *War on the Rocks*, March 9, 2018, https://warontherocks.com/2018/03/the-sunset-of-the-predator-reflections-on-the-end-of-an-era.

Clark, Bryan, Daniel Patt, and Harrison Schramm. "Decision Maneuver: The Next Revolution in Military Affairs." *Over the Horizon Journal*, April 29, 2019, https://othjournal.com/2019/04/29/decision-maneuver-the-next-revolution-in-military-affairs/.

Cohen, Eliot A., and Thomas A. Keaney. *Gulf War Air Power Survey Summary Report* (Washington, DC: Department of the Air Force, 1993).

Colby, Elbridge. "How to Win America's Next War." *Foreign Policy*, Spring 2019.

———. Testimony Before the Senate Armed Services Committee. January 29, 2019, https://www.armed-services.senate.gov/download/colby_01-29-19.

Cuomo, Scott, Olivia Garard, Jeff Cummings, and Noah Spataro. "Not Yet

Openly at War, but Still Mostly at Peace." *Marine Corps Gazette,* February 2019.

Deptula, Lt. Gen. David A., Heather Penney, Maj. Gen. Lawrence A. Stutzriem, and Mark Gunzinger. *Restoring America's Military Competitiveness: Mosaic Warfare* (Arlington, VA: Mitchell Institute for Aerospace Studies, 2019).

Dixon, Norman. *The Psychology of Military Incompetence* (New York: Basic Books, 1976).

Domingos, Pedro. *The Master Algorithm: How the Quest for the Ultimate Learning Machine Will Remake Our World* (New York: Basic Books, 2015).

Dougherty, Christopher M. *Why America Needs a New Way of War* (Washington, DC: Center for a New American Security, 2019). https://s3.amazonaws.com/files.cnas.org/CNAS+Report+-+ANAWOW+-+FINAL2.pdf.

Dunlap, Charles. "The Moral Hazard of Inaction in War." *War on the Rocks,* August 19, 2016, https://warontherocks.com/2016/08/the-moral-hazard-of-inaction-in-war/

Eaglen, Mackenzie. "Just Say No: The Pentagon Needs to Drop the Distractions and Move Great Power Competition Beyond Lip Service." *War on the Rocks,* October 28, 2019, https://warontherocks.com/2019/10/just-say-no-the-pentagon-needs-to-drop-the-distractions-and-move-great-power-competition-beyond-lip-service.

Economy, Elizabeth C. *The Third Revolution: Xi Jinping and the New Chinese State* (New York: Oxford University Press, 2018).

Edelman, Eric, and Gary Roughead. *Providing for the Common Defense: The Assessment and Recommendations of the National Defense Strategy Commission* (Washington, DC: United States Institute of Peace, 2018). https://www.usip.org/sites/default/files/2018-11/providing-for-the-common-defense.pdf.

Engstrom, Jeffrey. *Systems Confrontation and Systems Destruction Warfare* (Santa Monica, CA: RAND Corporation, 2018).

Fravel, M. Taylor. *Active Defense: China's Military Strategy Since 1945* (Princeton, NJ: Princeton University Press, 2019).

Freedman, Lawrence. *The Future of War: A History* (New York: PublicAffairs, 2017).

Gallagher, Mike. "State of (Deterrence by) Denial." *Washington Quarterly* 42, no. 2 (Summer 2019).

Gompert, David C., Astrid Struth Cevallos, and Cristina L. Garafola. *War with China: Thinking Through the Unthinkable* (Santa Monica, CA: RAND Corporation, 2016).

Hammes, T. X. *The Melians Revenge.* Atlantic Council Issue Brief (Washington, DC: Atlantic Council, 2019).

———. "Technological Change and the Fourth Industrial Revolution." In *Beyond Disruption: Technology's Challenge to Governance,* edited by George P. Shultz, Jim Hoagland, and James Timbie (Palo Alto, CA: Hoover Press, 2018).

Hannas, Wm. C., and Huey-meei Chang. *China's Access to Foreign AI Technology: An Assessment* (Washington, DC: Center for Security and Emerging Technology, 2019).

Harold, Scott W. *Defeat, Not Merely Compete: China's Views of Its Military Aerospace Goals and Requirements in Relation to the United States* (Santa Monica, CA: RAND Corporation, 2018).

Harrison, Todd. *Defense Modernization Plans Through the 2020s: Addressing the Bow Wave* (Washington, DC: Center for Strategic and International Studies, 2016).

Heginbotham, Eric, and Jacob L. Heim. "Deterring Without Dominance: Discouraging Chinese Adventurism under Austerity." *Washington Quarterly,* Spring 2015, 185–199.

Hicks, Kathleen, Melissa Dalton, Alice Hunt Friend, Lindsey Shepherd, and Joseph Federici. *By Other Means: Part 2: Adapting to Compete in the Gray Zone* (Washington, DC: Center for Strategic and International Studies, 2019). https://csis-prod.s3.amazonaws.com/s3fs-public/publication/Hicks_GrayZone_II_interior_v8_PAGES.pdf.

Hicks, Kathleen, and Alice Hunt Friend. *By Other Means: Part 1: Campaigning in the Gray Zone* (Washington, DC: Center for Strategic and International Studies, 2019). https://csis-prod.s3.amazonaws.com/s3fs-public/publication/Hicks_GrayZone_interior_v4_FULL_WEB_0.pdf.

Hoffman, Frank. "The Hypocrisy of the Techno-Moralists in the Coming Age of Autonomy." *War on the Rocks,* March 6, 2019, https://warontherocks.com/2019/03/the-hypocrisy-of-the-techno-moralists-in-the-coming-age-of-autonomy.

———. "Squaring Clausewitz's Trinity in the Age of Autonomous Weapons." *Orbis,* Winter 2019, 44–63.

Horowitz, Michael. "The Algorithms of August." *Foreign Policy,* September 12, 2018, https://foreignpolicy.com/2018/09/12/will-the-united-states-lose-the-artificial-intelligence-arms-race.

———. "Artificial Intelligence, International Competition, and the Balance of Power." *Texas National Security Review* 1, no. 3 (May 2018).

Horowitz, Michael, Gregory C. Allen, Elsa B. Kania, and Paul Scharre. *Strategic Competition in an Era of Artificial Intelligence* (Washington, DC: Center for a New American Security, 2018).

Hunter, Andrew P., Samantha Cohen, Gregory Sanders, Samuel Mooney, and Marielle Roth. *New Entrants and Small Business Graduation in the Market for Federal Contracts* (Washington, DC: Center for Strategic and International Studies, 2018). https://csis-prod.s3.amazonaws.com/s3fs-public/publication/181120_NewEntrantsandSmallBusiness_WEB.pdf?GoT2hzpdiSBJXUyX.lMMoHHcrBrzzoEf.

Jackson, Van. "Competition with China Isn't a Strategy." *War on the Rocks*, October 5, 2018, https://warontherocks.com/2018/10/competition-with-china-isnt-a-strategy/.

———. "Toward a Progressive Theory of Security." *War on the Rocks*, December 6, 2018, https://warontherocks.com/2018/12/toward-a-progressive-theory-of-security/.

Johnson, Aaron M., and Sidney Axinn. "The Morality of Autonomous Robots." *Journal of Military Ethics* 12, no. 2 (2013).

Kania, Elsa B. "The AI Titans Security Dilemmas." Governance in an Emerging New World, Hoover Institution. October 29, 2018. https://www.hoover.org/research/ai-titans.

———. *Battlefield Singularity: Artificial Intelligence, Military Revolution, and China's Future Military Power* (Washington, DC: Center for a New American Security, 2017).

Kania, Elsa B., and John K. Costello. *Quantum Supremacy? China's Ambitions and the Challenge to U.S. Innovation Leadership* (Washington, DC: Center for a New American Security, 2018).

Karlin, Mara. "How to Read the 2018 National Defense Strategy." Brookings Institution. January 21, 2018. https://www.brookings.edu/blog/order-from-chaos/2018/01/21/how-to-read-the-2018-national-defense-strategy.

Knox, MacGregor, and Williamson Murray. *The Dynamics of Military Revolution, 1300–2050* (Cambridge: Cambridge University Press, 2001).

Krepinevich, Andrew F. *The Military-Technical Revolution: A Preliminary Assessment* (Washington, DC: Center for Strategic and Budgetary Assessments, 2002).

———. *Preserving the Balance: A U.S. Eurasia Defense Strategy* (Washington, DC: Center for Strategic and Budgetary Assessments, 2017).

Krepinevich, Andrew F., and Barry Watts. *The Last Warrior: Andrew Marshall and the Shaping of Modern American Defense Strategy* (New York: Basic Books, 2015).

Lee, Kai-Fu. *AI Superpowers: China, Silicon Valley, and the New World Order* (New York: Houghton Mifflin Harcourt, 2018).

Luttwak, Edward N. "Breaking the Bank." *American Interest* 3, no. 1 (2007).

MacDonald, Julia, and Jacquelyn Schneider. "Trust, Confidence, and the Future of War." *War on the Rocks*, February 5, 2018, https://warontherocks.com/2018/02/trust-confidence-future-warfare.

Mahnken, Thomas. *Technology and the American Way of War Since 1945* (Ithaca, NY: Cornell University Press, 2008).

———. "Weapons: The Growth & Spread of the Precision-Strike Regime." *Daedalus* 140, no. 3 (2011): 45–57, https://doi.org/10.1162/DAED_a_00097.

Manyika, James, and William H. McRaven. *Innovation and National Security: Keeping Our Edge.* Council on Foreign Relations Independent Task Force No. 77 (New York: Council on Foreign Relations, 2019).

McFate, Sean. *The New Rules of War: Victory in the Age of Durable Disorder* (New York: William Morrow, 2019).

Mitre, Jim. "A Eulogy for the Two-War Construct." *Washington Quarterly* 41, no. 4 (2018).

Montgomery, Mark, and Eric Sayers. "Addressing America's Operational Shortfall in the Pacific." *War on the Rocks*, June 18, 2019, https://warontherocks.com/2019/06/addressing-americas-operational-shortfall-in-the-pacific.

Murray, Williamson, and Allan R. Millet, eds. *Military Innovation in the Interwar Period* (New York: Cambridge University Press, 1996).

National Security Commission on Artificial Intelligence. *Interim Report* (Washington, DC: National Security Commission on Artificial Intelligence, November 2019). https://drive.google.com/file/d/153OrxnuGEjsUvlxWsFYauslwNeCEkvUb/view.

Ochmanek, David, Peter Wilson, Brenna Allen, John Speed Meyers, and Carter C. Price. *U.S. Military Capabilities and Forces for a Dangerous World* (Santa Monica, CA: RAND Corporation, 2017).

O'Hanlon, Michael. *A Retrospective on the So-Called Revolution in Military Affairs* (Washington, DC: Brookings Institution, 2018).

———. *Technological Change and the Future of Warfare* (Washington, DC: Brookings Institution, 2000).

O'Mara, Margaret. *The Code: Silicon Valley and the Remaking of America* (New York: Penguin Press, 2019).

Owens, William A., with Ed Offley. *Lifting the Fog of War* (Baltimore: Johns Hopkins University Press, 2001).

Pillsbury, Michael. *The Hundred-Year Marathon: China's Secret Strategy to Replace America as the Global Superpower* (New York: Griffin, 2016).

Posen, Barry. *The Sources of Military Doctrine: France, Britain, and Germany Between the World Wars* (Ithaca, NY: Cornell University Press, 1984).

Roff, Heather, with David Danks. "The Necessity and Limits of Trust in Autonomous Weapons System." *Journal of Military Ethics*, 2018.

Rosen, Stephen Peter. *Winning the Next War: Innovation and the Modern Military* (Ithaca, NY: Cornell University Press, 1991).

Sanger, David. *The Perfect Weapon: War, Sabotage, and Fear in the Cyber Age* (New York: Crown, 2018).

Scales, Robert. "The Great Duality and the Future of the Army: Does Technology Favor the Offensive or Defensive?" *War on the Rocks*, September 3, 2019. https://warontherocks.com/2019/09/the-great-duality-and-the-future-of-the-army-does-technology-favor-the-offensive-or-defensive.

Scharre, Paul. *Army of None: Autonomous Weapons and the Future of War* (New York: W. W. Norton, 2018).

———. "The Real Dangers of an AI Arms Race." *Foreign Affairs*, May/June 2019, https://www.foreignaffairs.com/articles/2019-04-16/killer-apps.

Schneider, Jacquelyn, with Julia Macdonald. "Why Troops Don't Trust Drones." *Foreign Affairs*, December 20, 2017, https://www.foreignaffairs.com/articles/united-states/2017-12-20/why-troops-dont-trust-drones.

Sheehan, Neil. *A Firey Peace in a Cold War* (New York: Vintage Books, 2009).

Sherman, Justin. "Reframing the U.S.-China AI 'Arms Race.'" *New America*, March 6, 2019, https://www.newamerica.org/cybersecurity-initiative/reports/essay-reframing-the-us-china-ai-arms-race.

Shlapak, David A., and Michael Johnson. *Reinforcing Deterrence on NATO's Eastern Flank: Wargaming the Defense of the Baltics* (Santa Monica, CA: RAND Corporation, 2016). https://www.rand.org/pubs/research_reports/RR1253.html.

Singer, P. W. *Wired for War: The Robotics Revolution and Conflict in the 21st Century* (New York: Penguin, 2009).

Singer, P. W., and August Cole. *Ghost Fleet: A Novel of the Next War* (New York: Mariner Books, 2015).

Swaine, Michael. "The Deepening U.S.-China Crisis: Origins and Solutions." Carnegie Endowment for International Peace, February 21, 2019. https://carnegieendowment.org/2019/02/21/deepening-u.s.-china-crisis-origins-and-solutions-pub-78429.

Tellis, Ashley J. "Pursuing Global Reach: China's Not So Long March Toward Preeminence." In *Strategic Asia 2019: China's Expanding Strategic Ambitions,* edited by Ashley J. Tellis, Alison Szalwinski, and Michael Wills (Washington, DC: National Bureau of Asian Research, 2019).

Townshend, Ashley, Brendan Thomas-Noone, with Matilda Steward. *Averting Crisis: American Strategy, Military Spending and Collective Defence in the Indo-Pacific* (Sydney: US Studies Centre at the University of Sydney, 2019).

United States Department of Defense. *Indo-Pacific Strategy Report* (Washington, DC: Department of Defense, 2019). https://media.defense.gov/2019/Jul/01/2002152311/-1/-1/1/DEPARTMENT-OF-DEFENSE-INDO-PACIFIC-STRATEGY-REPORT-2019.PDF.

———. *Report of the Quadrennial Defense Review* (Washington, DC: Department of Defense, 1997).

———. *Report of the Quadrennial Defense Review* (Washington, DC: Department of Defense, 2001).

———. *Summary of the 2018 National Defense Strategy of the United States of America* (Washington, DC: Department of Defense, 2018). https://dod.defense.gov/Portals/1/Documents/pubs/2018-National-Defense-Strategy-Summary.pdf.

United States Department of Defense, Defense Innovation Board. *AI Principles: Recommendations of the Ethical Use of Artificial Intelligence by the Department of Defense* (Washington, DC: Department of Defense, 2019). https://media.defense.gov/2019/Oct/31/2002204458/-1/-1/0/DIB_AI_PRINCIPLES_PRIMARY_DOCUMENT.PDF.

Vickers, Michael, and Robert Martinage. *Future Warfare 20XX Wargame Series: Lessons Learned Report* (Washington, DC: Center for Strategic and Budgetary Assessments, December 2001).

Work, Robert O., and Greg Grant. *Beating the Americans at Their Own Game: An Offset Strategy with Chinese Characteristics* (Washington, DC: Center for a New American Security, 2019).

Zupan, Daniel S. *War, Morality, and Autonomy: An Investigation of Just War Theory* (London: Routledge, 2017).

INDEX